Milk production: Science and practice

D1744540

Longman Handbooks in Agriculture
Series editors

C. T. Whittemore
K. Simpson

Milk production: science and practice

J.D. Leaver

Farm Director, Crichton Royal Farm,
The West of Scotland Agricultural College

Longman
Scientific &
Technical

Longman Scientific & Technical
Longman Group UK Limited
Longman House, Burnt Mill, Harlow
Essex CM20 2JE, England
and Associated Companies throughout the world

First published 1983
Reprinted by Longman Scientific & Technical *1987*

Set in 10/11pt Linotron 202 Times Roman
Produced by Longman Group (FE) Ltd
Printed in Hong Kong

British Library Cataloguing in Publication Data
Leaver, J. D.
  Milk production. – (Longman handbooks in
  agriculture)
  1. Dairying
  I. Title
  637′.1   SF239
  ISBN 0-582-44010-6

Library of Congress Cataloging in Publication Data
Leaver, J. D. (John David), 1942–
  Milk production.

  (Longman handbooks in agriculture)
  Bibliography: p
  Includes index.
  1. Dairy farming. 2. Dairying. 3. Milk.
I. Title. II. Series.
SF239.L427 1983   636.2′142   82-17925
ISBN 0-582-44010-6

ISBN 0-582-44010-6

Contents

Preface

The dairy enterprise provides the farmer not only with a challenging business, but also with a satisfying although arduous way of life. Modern systems of milk production have evolved through the rapid uptake of new technology, particularly in the last two decades, and a rudimentary knowledge of the science of milk production is now becoming a prerequisite for any farmer wishing to have a dairy enterprise which is profitable.

The development of a profitable enterprise depends on the business acumen of the farmer in using the correct balance of the available resources (land, labour and capital), and also on the use of good husbandry practices by adopting systems of management which are economically efficient.

The subject of milk production therefore presents a fascinating study of both biological and economic efficiency for enthusiasts at all levels of involvement, and the objectives of this handbook are to give the reader some insight into the science and practice of milk production.

There are many friends and associates who have contributed indirectly to this handbook through their influence and encouragement over many years. To them I owe a great debt. I would particularly like to thank Mrs June McCallay for typing many drafts of chapters with such good grace during the writing of the handbook.

J. D. Leaver
Dumfries 1982

The milk production industry

1

The production of milk by dairy farmers is not only a business but also a way of life. Many of the decisions taken on the farm are consequently dictated both by financial constraints and by the desire for job satisfaction. Nevertheless during recent decades there has been a continuing squeeze on profits in the milk production industry as prices have failed to keep pace with the increase in costs and the dramatic trends which have taken place in systems of milk production are mainly a response to reduced profits per cow. This has given the dairy farmer an increasing awareness of the need not only to produce milk more efficiently but also to have a stronger say in its marketing.

Trends in milk production

Dairy herd sizes

The number of dairy cows in the UK has remained fairly stable in recent years, but the number of milk producers has shown a continuous decline. The size of dairy herds thus increased from 20 cows in 1960 to 52 in 1979 (Fig. 1.1). In England and Wales in 1979, 33 per cent of cows were in herds of over 100 cows, and in Scotland where herd sizes are bigger, 43 per cent were in herds over 100 cows.

The increase in herd size has arisen partly from the desire for greater farm profits through increased scale, and partly

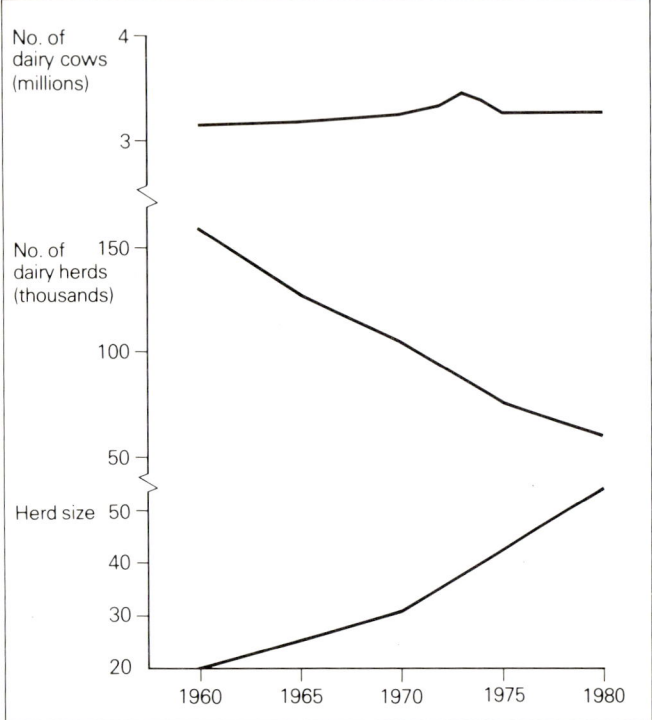

Figure 1.1   Changes in dairy herd structure between 1960 and 1980 (from *Dairy Facts and Figures 1980*)

from the reduced profit per cow. The trend towards larger dairy herds is likely to slow down in the late 1980s with a large proportion of herds being in the range 100–150 cows.

The majority of milk producers are situated in the wetter western half of the country where farms are smaller and where the systems of dairying rely to a large extent on grass and its conserved products. Dairying is ideally suited to the farm structure in the west due to its high and consistent level of profit per hectare compared with other ruminant livestock enterprises. In the drier eastern side of the country farms are larger and a wider range of enterprise options is available.

Milk yield and composition

Many of the changes in annual milk yields and composition have arisen from the change in breed structure as larger dairy breeds have replaced smaller breeds. Between 1970 and 1980 the Ayrshire breed in England and Wales declined from 10 to 3 per cent and in Scotland from almost 70 per cent down to 30 per cent of total cows. The Friesian breed and its crosses now account for over 90 per cent of cattle in England and Wales and 60 per cent in Scotland. The further introduction of the Holstein breed in the 1970s led to an increase in the size of the Friesian cow and to higher yields. Average yields of the major breeds are shown in Table 1.1.

The use of artificial insemination (AI) and nominated

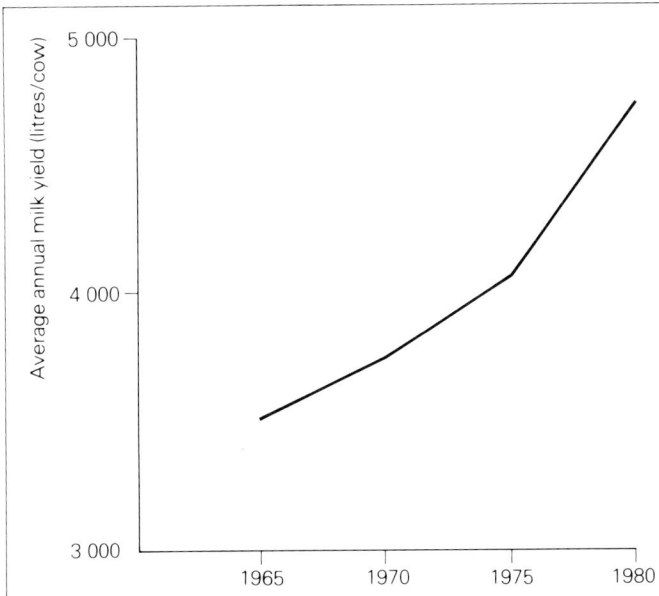

Figure 1.2   Average annual milk yield per cow in the UK

Table 1.1   Milk yields and butterfat of major breeds in recorded herds in 1978/79

Breed	Average yield (kg)	Average butterfat (%)
Friesian	5 441	3.79
British Canadian Holstein	6 218	3.71
Ayrshire	4 863	3.94
Jersey	3 776	5.14
Guernsey	3 941	4.65
Dairy Shorthorn	4 730	3.62

Source: Dairy Facts and Figures 1980

progeny-tested sires and better feeding management have also contributed to milk yield increases which have averaged over 2 per cent per year (Fig. 1.2).

Housing and milking systems

Up to 1950 almost all dairy cows were housed in cowsheds (also called byres and shippons) in which they were also milked. The straw-yard loose housing system combined with parlour milking was introduced in the next decade and this had a great advantage in labour saving for milking, bedding and cleaning. However, the amount and cost of straw bedding for this system for farms in non-arable areas led to the development of cubicle housing in the 1960s, and over 60 per cent of herds are now housed in this way.

Unfortunately with cubicle housing the dung and urine are mixed together to form a slurry, and methods of storing and spreading slurry have contributed to increased costs.

When the cowshed declined in the 1960s various milking parlour systems were developed to cope with increasing cow numbers and faster rates of milking. The early tandem, chute and abreast parlours gave way to the herringbone which remains the most popular parlour for herds up to 200 cows. For larger herds rotary parlours have been developed with throughputs of over 120 cows/man hour, although these have the disadvantages of high capital and maintenance costs.

Feeding systems

The major change in feeding dairy cows has been the swing from hay to silage as the main conserved forage. Very few herds of over 75 cows now feed hay, and traditional feeds such as kale, swedes and turnips have also declined in importance. For herds of over 100 cows, carting of silage by fore-end loader or by forage box to be fed in passageways has become common practice. For smaller herds, self-feeding of silage is likely to remain an efficient low-cost system.

The use of compound concentrates has increased in line with increasing milk yields, and the methods of feeding have also changed. Higher levels of feeding have led to systems of feeding concentrates in the housing area (out-of-parlour feeders) making use of electronics to feed programmed amounts to individual cows. A further development is to mix the concentrates with the forage in a mixer wagon (complete diets). The high capital costs of the latter systems are likely to limit their use.

Milk production as a business

The management of the dairy herd requires a host of decisions to be made both in the general organization and the day to day running of the farm. A whole range of options which will provide an acceptable level of profitability are feasible in theory, but in practice each farm has particular resources of land, labour and capital and these together with the aptitude of the farmer will be the major determinants of the system of milk production chosen.

Resources available

Land is an important factor dictating the type of milk production system. Two extreme situations are where the land of the farm supplies all the dairy cows' feed and where, as in the case of the original town dairies, all the feed is purchased. Most dairy farms, however, lie somewhere

between these extremes, with the forage and grazing being
provided by the farm with some or all of the concentrate
feed being purchased.

The land required per cow can be reduced in two ways:
1. Increase land productivity (more fertilizer, better
drainage and so on).
2. Increase the purchase of feeds (forage and/or
concentrates).

The number of cows carried on the farm can also be
increased by changes in youngstock management. Reducing
the age at first calving of reared heifers from an average of
33 months to 24 months, or not rearing on the farm at all
releases hectares of land which allows the herd size to be
increased (Table 1.2).

A large proportion of dairy farms are family farms which
employ no hired labour. The need for increased
productivity per man may not seem as important, therefore
as when there is hired labour. The use of labour-saving
systems, such as efficient milking parlours will thus depend
on the state of the overdraft relative to the profitability of
the herd, and the number of hours that the family is
prepared to work. Where labour is hired, good wages have
to be paid to attract a person who is prepared to take on
considerable responsibility, and to work long and unsocial
hours without supervision. Consequently simplified systems
of housing and automation in milking and feeding have
been developed to allow each man to look after more cows,

Table 1.2 Effect of rearing policy on herd size

	Extensive system replacements calving at 3 yrs	Intensive system replacements calving at 2 yrs	Intensive system with no replacements
Land required (ha)			
per cow	0.60	0.40	0.40
per replacement unit	0.25	0.10	–
Total	0.85	0.50	0.40
Number of dairy cows per 50 ha	59	100	125

and it is not unusual to have a ratio of one man per 60–80
cows.

Dairying is an enterprise with a high capital requirement
in cows, youngstock, housing and machinery. If the farm is
owner-occupied there is also considerable capital tied up in
land. Working capital requirements for concentrates,
fertilizer, paid labour, for example, are also high, although
this is offset to some extent by the regular monthly payment
for milk, which does not occur with most other livestock
enterprises.

Capital investments thus require careful study before
implementation, particularly when interest rates are high.

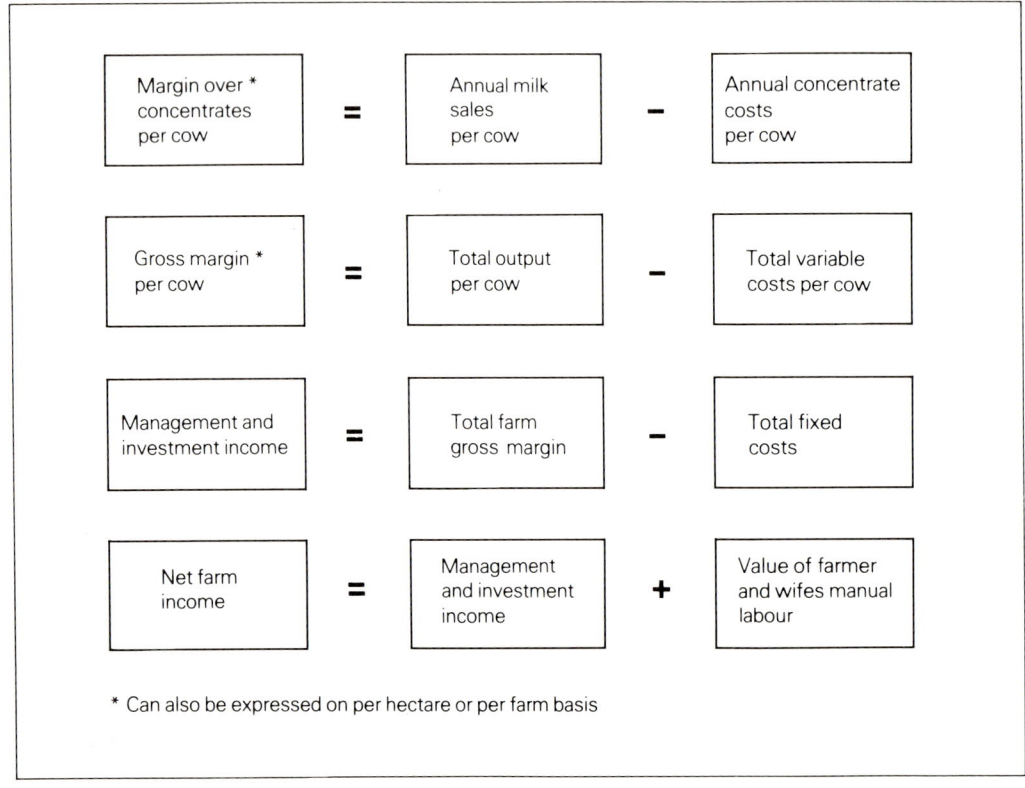

* Can also be expressed on per hectare or per farm basis

Figure 1.3 Some measures of the profitability of milk production

The proportion of the capital in the business which is borrowed is called the 'gearing', and although a farm with high gearing is very profitable when the business is running efficiently, it is more susceptible to inefficiencies and to external changes such as an increase in interest rates. The conservative nature of farmers has meant that farms have traditionally had low gearing, but financial constraints have increasingly forced dairy farmers to increase their borrowings.

Measures of profitability

Some commonly used measurements of profitability are shown in Fig. 1.3.

Margin over concentrates per cow is the simplest measure and is used in most farm costing surveys, but means little in comparing different farms mainly because it does not take into account the stocking rate. It is, however, a simple, easily calculated index of the trends within individual farms. The margin over concentrates per hectare is a more satisfactory measure of the management efficiency of the farm.

The gross margin which includes all variable costs (purchased feeds, fertilizer, seeds, veterinary etc.) is commonly used to compare enterprises. For the individual farmer a more important measure of profit is the management and investment income, which is the reward for the management input, and which provides the return for capital invested in the farm. This sum has to meet charges such as interest and repayments on borrowed money, income tax and must provide for future investment. The net farm income which additionally includes a value for the farmer and wife's manual labour represents the sum of money on which the farmer has to live.

Factors associated with profitability

The three main elements which explain differences between farms in gross margin per hectare are:
1. Annual milk sales (litres/cow).
2. Annual stocking rate (cows/ha).
3. Annual purchased feeds (t/cow).

The relationship of these factors to gross margins for costed farms are illustrated in Fig. 1.4.

When the performances of a large number of farms are plotted out, milk sales per cow and stocking rate are strongly correlated with gross margin per hectare, but purchased feeds have only a weak relationship. This is because at all levels of input there are farms with both high and low gross margins, i.e. concentrates are used with a wide range of efficiency. Also there is a strong interaction of purchased feed input with both milk yield and stocking rate, and it is profitable to feed high levels of concentrate if milk yields and stocking rate are high.

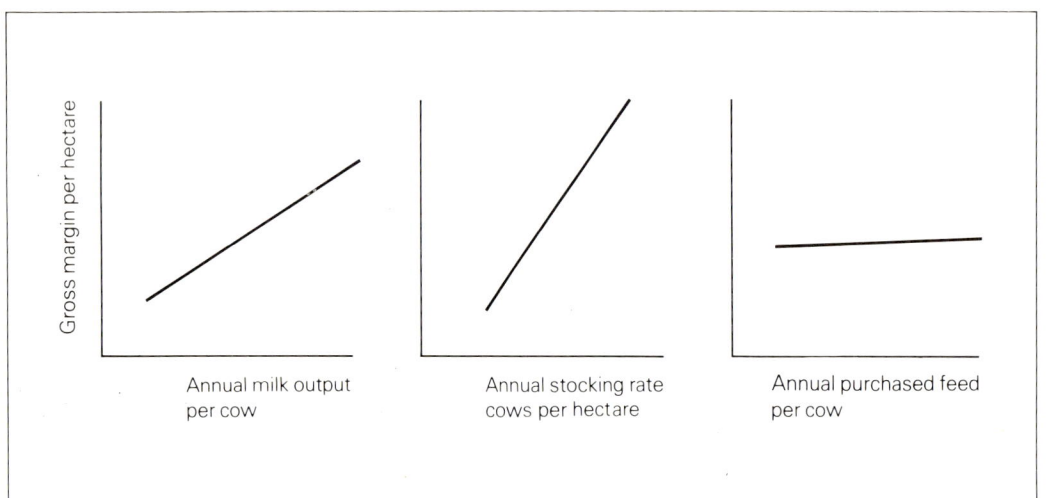

Figure 1.4 The relationship of three important parameters with the gross margin per hectare

High or low concentrate system

The trends in dairying are towards intensification with increased milk outputs per hectare and much of this is brought about by greater inputs of concentrates which increase milk yield per cow and allow stocking rates to be increased.

The average response to additional concentrates where cows are fed *ad libitum* forage is about 1 kg milk/additional 1 kg of concentrates. The reason that the concentrate required is greater than 0.45 kg, which represents the energy requirement, is mainly because the concentrates substitute for forage. Increasing the concentrate input thus allows stocking rates to be increased, as shown in Table 1.3.

Table 1.3 Effect of concentrate input on milk output, stocking rate, and margin over concentrates (M/C) at the same level of grass utilization

Annual milk yield (kg/cow)	Annual concentrates (t/cow)	Annual stocking rate (cows/ha)	Relative milk output per ha	Relative M/C per cow	Relative M/C per ha
4 000	0	1.65	100	100	100
5 000	1	1.89	143	98	112
6 000	2	2.21	201	96	129
7 000	3	2.67	283	94	153

The four example systems all have the same level of grass utilization (7 t dry matter (DM)/ha). Increasing the concentrate levels thus gives rise to large increases in milk output per hectare. The effect on margin over concentrates per cow is negligible but the margin per hectare is substantially increased. The high concentrate approach, however, requires greater fixed costs associated with keeping more cows, and whether this is advantageous will depend on the size of the farm, whether it is tenanted or owner-occupied, and the gearing of the business. A non-financial consideration but one which has high priority is whether the farmer obtains his job satisfaction from achieving high yields per cow, or from achieving a high level of self-sufficiency from the farm; in other words, producing milk mainly from grass and conserved forage.

Seasonality of production

The price of milk paid to the farmer by the Milk Boards varies according to the season, with the highest prices being paid in the winter months when milk is more expensive to produce. This is to ensure a more regular supply of milk throughout the year.

Cows calving in the winter months have a much 'flatter' lactation curve than those calving in the spring (Fig. 1.5).

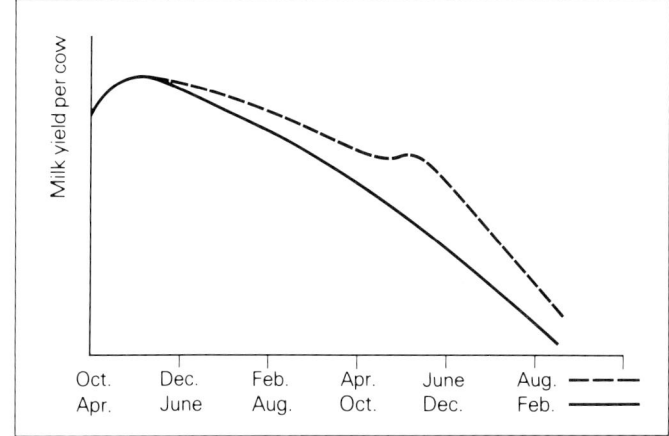

Figure 1.5 Lactation curves for October – and April – calving cows of similar genetic potential

The latter show a greater rate of decline in yield, due to the declining availability and quality of grazed grass during the season and this results in shorter lactation lengths and lower lactation yields.

There are very few herds which practise block calving, i.e. all the cows calve in a 12-week period or less. This system has advantages in management but requires a strict discipline in breeding management which few farmers are willing to accept. Consequently most farmers calve about 70 per cent of the herd between August and December, and 30 per cent between January and April. Other than in the south-west, autumn/winter calving is generally more profitable than spring calving because it forces farmers into buying more feed which allows him to keep more cows.

Milk marketing

The Milk Boards

In the UK the marketing of milk has been controlled by the Milk Marketing Boards in England, Wales, Scotland and Northern Ireland since 1933. The Boards are producer organizations which have the exclusive right to buy all milk from producers, and to equalize prices paid to producers (the pool price) irrespective of the use to which the milk is intended.

The activities of the Boards are therefore mainly concerned with buying milk, transporting it to dairies and creameries, and marketing milk and its products. In addition they provide technical services to the producer such as the AI organizations and farm management services.

Milk pricing

The income of the Milk Boards (and therefore the prices they are able to pay the milk producers) is controlled by the value of sales of milk for liquid consumption and for manufacture. Control over the price of milk in the liquid market is the responsibility of the government, and the support for the dairy industry is provided by the EEC intervention system of buying butter and skimmed milk powder.

Each Milk Board publishes a provisional schedule of monthly prices to producers at the start of each year (April). These are based on estimates of the returns from liquid and manufacturing markets, and on estimates of the costs of running the Board's activities. The actual monthly price paid to producers is confirmed or adjusted each month.

Additions or subtractions are made to the monthly price of each producer on the basis of the compositional quality of the milk. The scheme of payment for each Board is different, being based on fat protein, lactose, solids-not-fat

(SNF) content and/or on total solids content. Other deductions are made for hygiene failure, antibiotic contamination, and also for the co-responsibility levy which is the EEC scheme to assist in financing the disposal of surpluses of milk and dairy products.

Milk utilization

Total milk production in the UK in 1979/80 was 15 160 million litres of which 48 per cent was sold on the liquid market, the remainder being used for manufacture. Estimated liquid consumption of milk is about 2.6 litres per person per week. The distribution of manufactured milk into different products is shown in Table 1.4.

Table 1.4   Utilisation of milk for manufacture in the UK in 1979/1980

Product	(%)
Butter	46
Cheese	30
Condensed milk	7
Wholemilk powder	3
Fresh cream	12
Sterilized cream	1
Other products	1
	100

Source: Dairy Facts and Figures 1980

European Economic Community

After the transitional period into the EEC, the UK system of guaranteed prices was terminated at the end of 1977. The EEC policy for dairying is to manage the Community Markets to secure product prices which permit milk producers to obtain the target price for milk. This is determined by the Council of Agricultural Ministers and is not a guaranteed price.

The methods used to secure appropriate product prices are:
1. The operation of an intervention system for butter and skimmed milk powder, where agencies are obliged to buy products offered to them at pre-determined prices (about 95 per cent of target price) which allow surpluses to be temporarily taken off the market.
2. The imposition of variable levies on imports from non-EEC countries.
3. The payment of subsidies on surplus products, e.g. skimmed milk powder fed to stock and butter for domestic consumption.
4. The payment of subsidies on exports to allow EEC produce to compete with the lower prices on international markets.

Prices, subsidies and levies are fixed in terms of the European Currency Unit (ECU). These units are converted to the currencies of each member state by special rates of exchange known as 'green rates'.

Genetic improvement

2

The transition of the early bovine species to the modern dairy and beef breeds has taken many centuries. Following the early domestication of wild cattle for draft purposes, there was subsequent selection for milking ability and fat deposition, and the two main types of domestic cattle which have evolved are *Bos indicus* which is found in tropical countries and *Bos taurus* found in the more temperate regions.

Great improvements in cattle performance were not seen, however, until the eighteenth century, when the pioneering work of animal breeders such as Robert Bakewell led to the uptake of objective breeding methods. This was followed up in the nineteenth century by the emergence of Breed Societies. These societies laid down objectives relating to the type of cattle to be selected for, and this led both to a greater uniformity of type and to an improved level of performance.

The lactational performance of the national dairy herd continues to increase annually (Fig. 1.2, p. 3), due partly to genetic improvement and partly to better management. Much of this improved performance arises from dairy farmers with herds at the lower end of the production scale going out of business, and it is therefore difficult to assess the relative contributions of genetics and management to the overall improvement. However, compared with only 20–30 years ago, the modern dairy cow is a highly selected individual with the ability, in many cases, to produce over

8 000 kg of milk in a year, and with the body conformation to support this yield over a number of lactations.

Principles

The basic method of genetic improvement is selection, which means choosing the parents of the next generation, so that the genetic merit or genotype of the next generation is improved in the desired traits. The genotype represents the maximum performance that the cow could possibly achieve, and environmental factors concerned with management determine how close it actually comes to achieving this potential. The performance that is achieved on the farm, e.g. milk yield, is called the phenotype, and represents the combined effects of genotype and environment.

The traits which are selected for may be either qualitative or quantitative. With qualitative traits the cow either does or does not possess the particular trait such as horns or hair colour. These traits are controlled by only a few genes. Quantitative traits include most of those concerned with milk production. These are controlled by a large number of genes, and therefore progress in these traits is made by increasing the proportion of desirable genes at the expense of undesirable genes. For example, the genes controlling milk yield may be concerned with udder size, the amount of secretory tissue, blood supply to the udder, the digestion and metabolism of food, food intake and many other characteristics. It is therefore not surprising that only a proportion of cows possess all the necessary gene combinations for high milk yields.

Genetic improvement in production traits is therefore concerned with probabilities. Mating an outstanding sire for milk yield with an outstanding dam does not guarantee that the resulting offspring will also be outstanding, although the probability of this occurring is greater than if the sire or dam were of low genetic merit. The spermatozoa in the semen of a bull are not identical, neither are the ova produced by a cow, because a great variety of combinations of genes can be produced during meiosis (the division of the male and female germ cells to produce spermatozoa and ova respectively). This can be illustrated by the performance of unidentical (dizygous) and identical (monozygous) twin cattle. Unidentical twins, which are produced when two separate ova are fertilized by two separate spermatozoa, can be quite different in their subsequent performance even though they have the same sire and dam. However, identical twins, which are produced when a fertilized ovum divides to produce two identical genotypes, have a very similar performance if maintained in the same environmental conditions.

The rate at which a trait is improved depends on its heritability, and on the relationship (positive or negative) between this trait and other traits which are being selected for. A list of heritability estimates and the correlation of individual traits with milk yield are given in Table 2.1. The phenotype of a cow (its measured performance) is a result of a combination of genetic and environmental influences, and the heritability gives a measure of that proportion of the phenotypic variation which is due to genetic influences.

Table 2.1 Heritability of traits and their correlation with milk yield

Trait	Heritability	Correlation with milk yield	
		Phenotypic	Genotypic
Milk yield	0.25		
Fat yield	0.25	+0.8	+0.8
Total solids yield	0.25	+0.9	+0.9
Fat %	0.50	−0.3	−0.4
SNF %	0.50	−0.3	−0.4
Body size	0.40	+0.2	+0.1
Reproductive efficiency	0.05	+0.1	0
Longevity	0.05	+0.3	+0.7
Milking ability	0.30	+0.1	+0.1
Type	0.25	−0.1	−0.2

The practical consequences for the cattle breeder are that traits with a high heritability can be improved more quickly by selection than traits with a low heritability. The correlation between traits signifies whether improvemement in the trait selected for will increase or decrease the performance in other traits, and to what extent. If two desirable traits are highly positively correlated then only one of the traits needs to be selected for as genetic improvement in the other trait will automatically follow. If there is no correlation between two traits, selection for one trait will have little effect on the other trait, but if they are negatively correlated, selection for one will reduce the performance in the other. From Table 2.1 it is clear, for example, that selection for milk yield will result in a slower rate of genetic progress than selection for milk fat percentage. Also, selection for type (body conformation in relation to Breed Society preferences) will result in a slight reduction in genetic merit for milk yield.

Objectives

The most important considerations in choosing traits for genetic improvement are:
1. The trait must be measurable. In the dairy cow the traits concerned with milk production (milk yield and composition) can be measured precisely in individual cows. Traits concerned with frequencies of occurrence, such as disease, can be measured if a satisfactory recording system is practised, but large numbers of cattle are required to obtain an accurate measure. The

measurement of traits concerned with type are more subjective and the problem in setting out objectives for this trait is that there is not always a consensus of opinion on what is desirable.

2. The trait must be marketable. For most dairy farmers the major income is derived from the sale of milk solids, with additional income from the sale of cull cows and from the value of calves. The objective of commercial producers must therefore be to increase the yield of milk solids. For a small minority of dairy farmers producing pedigree cattle for sale, there is a substantial income from the sale of breeding stock, and attention to type will be an important breeding objective.
3. The trait should have a moderate to high heritability. This ensures that progress in genetic merit for selected traits can be made at a reasonable rate.
4. The trait should be positively correlated with other desirable traits. If a negative relationship exists with other desirable traits, selection for the trait will be at their expense. This is particularly the case if there is a high negative correlation.
5. It should be practicable to record the trait both objectively and accurately. Fortunately in the cow the most important traits, milk yield and composition can be measured reasonably easily and precisely.

Traits for selection

Milk production

High milk yield is the trait often assumed to be related to profitability. However, the growing trend towards milk payments based on some aspects of compositional quality has emphasized the need to observe closely the fat and solids-not-fat (SNF) or protein content of the milk.

The correlations between milk yield and milk solids content are negative, and therefore the major trait to be selected for should be the yield of milk solids. This has a heritability of 0.25 and therefore a moderate rate of progress can be made in selecting for this trait.

Type

The evaluation of dairy cows on the basis of their body conformation has been carried out for centuries, but the development of Breed Societies and the popularity of showing cattle has increased the emphasis on type (Fig. 2.1).

The original rating of cows by the Breed Societies in terms of Excellent, Very Good, Good Plus, Good, Fair and Poor has been supplemented in recent years by more complex systems giving independent ratings to different aspects of conformation such as body capacity, dairy character, mammary system, legs and feet, general

(a)

(b)

Figure 2.1   British Friesian cows which have been type classified as having: (a) very good; (b) good plus; (c) good; and (d) poor conformation

(c)

(d)

appearance and temperament.

The major problems associated with type classification are that some of the individual traits have no relationship with productivity or profitability and therefore are of little value to the commercial producer. Also there may be differences between individual classifiers in their judgement of type, and animals may rate differently at different stages of lactation.

It is often claimed that Excellent and Very Good cows have greater longevity than those with a lower type-classification. However, in those herds where type is considered to be a desirable trait, there is a greater tendency to retain in the herd those cows with a high type-classification, so that inevitably they build up a high lifetime performance even though their lactation yields may not be outstanding. Conversely, cows with similar yields but poorer type are more likely to be culled earlier in these herds. The result is that in many pedigree herds there is a contrived relationship between type and longevity.

The reasons for culling cows from the herd are mainly (75 per cent) accounted for by low milk yield, reproductive problems and mastitis (Table 2.2). Selection for type will thus have little effect on the incidence of these culls and inevitably will have little effect on longevity.

In general, cows with high first lactation yields tend to remain in the herd longer than those with lower first lactation yields, and therefore lifetime production is closely correlated with genetic merit for milk production. The correlations between milk production and type traits are low and in some cases negative and so selection for both traits in a herd will result in only a slow genetic improvement in both. The farmer must decide the relative importance of milk production and type in his dairy enterprise. It is difficult to evaluate their relative economic worth – to the commercial producer, milk production should be the main objective, whereas the pedigree breeder selling breeding stock will inevitably place more emphasis on type.

Table 2.2 Reasons for disposal of dairy cows

	Culls (%)
Reproductive problems	35
Low milk yield	25
Mastitis	15
Metabolic disorders	5
Accident/deaths	3
Milking problems	5
Legs/feet	5
Others	7

Milking ability

The continuing increase in herd size has meant that a

greater proportion of the stockman's time is spent milking cows. Parlours have been designed to allow a quicker throughput of cows per man hour, but in many cases slow-milking cows can mean that the man is waiting for cows to milk out. Milking-out rate is therefore an increasingly important trait.

The rate of milking can be measured as the average flow rate (milk yield divided by milking out time), the peak flow rate (maximum yield in any one minute), or the percentage of the milk yield in the first 2 minutes. The heritabilities of these traits are moderately high (0.3–0.5) as the rate of milking depends mainly on the teat shape, size of teat orifice, and the teat sphincter. High yielding cows tend to be faster milkers, so although direct selection for milking ability might take place through the culling of slow milking cows, selection for milk production alone should also result in faster milking cows.

Body size

There has been a general move towards larger types of cows, for example a move from the Ayrshire breed to the Friesian and the Holstein, a major reason being that bigger cows produce more milk; however, they also have a greater maintenance requirement, and for each 1 kg increase in liveweight an extra 6–7 kg additional milk per lactation will be required to offset the cost of the additional energy requirements for maintenance. It can be argued, however, that the fixed costs (overheads) are lower per unit of milk output for a larger cow because the time spent in milking, feeding and cleaning are similar for big and small cows.

For many dairy farmers body size is not an important trait although it can be selected for quite easily as it has a moderately high heritability (0.4). It is positively correlated with milk yield and so there will be a tendency to select also for increased body size if milk yield is a main selection trait.

Reproductive efficiency

The heritability of traits associated with fertility is low. Calving interval, services per conception and calving to conception interval have heritabilities of less than 0.1, and there is little scope for genetic improvement through selection.

Some traits associated with specific infertility conditions such as cystic ovaries are more highly heritable. However, culling on the female side due to failure to conceive is an efficient natural selection mechanism of limiting the incidence of these specific problems.

The incidence of calving problems (dystokia), can be high in first calving heifers, and is important not only because a stillbirth might result, but also because the subsequent reproductive and lactational performance of the heifer is often adversely affected. The incidence of dystokia depends on the genetic make-up both of the heifer and of the sire of the calf. Selection of a bull which produces smaller calves is the most efficient method of preventing dystokia problems, although one disadvantage is that calves which are small due to their genetic make-up, often develop into small cows, and this may be undesirable.

Mastitis

Mastitis is a complex disease caused by a variety of organisms which gain entry to the udder. Breeding for resistance to mastitis therefore involves producing cows which have a mammary system which prevents entry of these organisms or which has a greater ability to control organisms which have entered the udder.

The genetic make-up of the cow does influence the size and shape of the udder, and the teat sphincter which is the main barrier to the entry of organisms to the udder. Heritability estimates for resistance to mastitis vary between 0.05 and 0.40, so in theory selection may be possible. However as with all diseases, one is concerned with measuring frequencies of occurrence and for mastitis in each lactation this will range from 0 to about 3 for most cows. If, for example, cows are culled which have more than three infections per lactation, then some selection is taking place on the female side. However, it is more difficult to collect the necessary information from a large number of daughters in order to select bulls for mastitis resistance.

Male selection

Progeny testing

The scope for culling on milk production traits on the female side is very limited in most herds and therefore genetic progress depends almost entirely on the use of superior bulls. The progeny testing of bulls is concerned with assessing the genetic merit of the daughters of one bull relative to that of other bulls. Because bulls used for artificial insemination (AI) purposes produce daughters in a number of herds, and because their yields in any one herd are a combined effect of genetics and management, it is necessary to compare their performance with daughters from other bulls within herds. This is termed a 'contemporary comparison'.

In a contemporary comparison, the first lactation yields of the daughters from a particular bull are compared with the yields of the daughters from other bulls, milked in the same herd and under the same management. A common criticism of progeny testing is that bulls are evaluated on the basis of the first lactation yields of their daughters, and not on their lifetime performance. A number of studies have shown, however, that daughters with high first lactation yields also stay in the herd for a longer period and therefore also have the better lifetime performances. Statistical procedures are used to give an estimate of the bull's genetic merit which in the UK is known as the 'Improved Contemporary

Comparison' (ICC), which estimates the difference between the performance of the daughters of the bull, and the daughters of a bull with an ICC of 0.

The ICC for a bull is calculated from the performance of his daughters in a number of herds. It predicts the average merit of his daughters for various milk production traits relative to a common baseline for the breed. The ICCs are expressed as:

± kg milk
± kg fat
± kg protein
± % fat
± % protein

They represent the best average predictions based on the information on daughters currently available. In calculating the ICC, adjustments are made for factors other than the genetic merit of the bull which can affect the milk production traits. These are:
1. The calving season.
2. The age at calving of heifers.
3. The genetic merit of the sires of the contemporaries.
4. The genetic merit of the sire of the dam.
5. The year group of the bull.

The ICC is more reliable the larger the number of daughters recorded, and the greater the number of herds in which these daughters are milked. In an ICC, the 'weighting' gives a measure of the reliability. This is often called 'the number of effective daughters' and represents the number of effective comparisons of the bull's progeny. ICCs based on weightings of less than 30 can only be regarded as provisional.

Progeny tests also provide quantitative estimates of the conformation (type) of a bull's daughters and this is a useful adjunct to the milk production results.

In selecting semen from a bull for use in a herd the following points should be considered:
1 Select a bull with an AI proof, a weighting of over 30 with daughters in more than 10 herds in one calving season.
2. The bull should have a high ICC for kg fat plus kg protein. Selection of a bull on milk yield alone can lead to a reduction in milk composition due to their negative correlation.
3. The average fat percentage should be 3.9 per cent or over to avoid payment penalties on the milk composition of the daughters.
4. Bulls with low conformation scores in traits relating to production (mammary system, legs and feet) should be avoided.

A problem in estimating the benefit to be derived from selecting a particular bull is that the plus (or minus) rating in a particular trait relates to a common baseline, and so a bull of +30 kg of fat plus protein may be a 0 kg bull in a herd of high genetic merit and +60 kg in a herd of low genetic merit. The benefits from using the bull can only be estimated, therefore, if the genetic merit of the herd is known, but as management is such an important factor in milk production it is difficult to measure. In herds where breeding is predominantly by AI, the genetic merit of the herd is estimated from the ICCs of the sires and the production indices of the dams of the cows. An important consequence of the herd merit is that the bull selected should have a genetic merit (ICC) at least equal to the genetic merit of the herd, otherwise the merit of the herd will decline (Table 2.3).

Table 2.3 Effect of the bull ICC and herd genetic merit on the genetic merit of daughters for the yield of milk fat + milk protein

Herd genetic merit (kg)	Bull ICC (kg)	Genetic merit of progeny (kg)
0	0	0
0	+30	+30
+30	0	+15
+30	+30	+45
+60	0	+30
+60	+30	+60

The major disadvantages of progeny testing are concerned with the cost of keeping bulls whilst awaiting the results of their progeny test. It is normal to select young bulls from matings of high genetic merit cows with progeny-tested bulls, and 750 straws of semen are taken from each bull for the initial progeny test. The young bull is then laid off until the results are known and this will normally take 4–5 years. The proportion of young progeny tested bulls which subsequently enter the AI stud is only 10–20 per cent, which illustrates that a high proportion of bulls with good pedigrees do not have a high genetic merit.

Pedigree selection

The pedigree of a bull gives a record of his close ancestors, and on its own it is of little value unless the productivity of these ancestors is known. Even then the pedigree only gives an indication of the bull's genetic merit.

For a trait of moderately high heritability of 0.4, e.g. body size, the correlation between a parent's breeding value and that of the progeny is only about 0.3, and between a grandparent's breeding value and progeny is only 0.1. At lower heritabilities, such as for milk yield (heritability 0.25), the respective correlations are only about 0.2 and 0.1.

The practice of using unproven bulls persists in many herds in spite of the detrimental effect on genetic progress in productive traits. Selection in these herds is carried out on the female side through the sale of surplus heifers and cows (particularly in pedigree herds) and through the culling of low-yielders. This practice results in an acceptable level of milk yield under good management conditions combined with a better standard of conformation. In economic terms it can only be justified if a large proportion of the income is derived from the sale of breeding stock. As each parent only contributes half of its genes to its offspring there is considerable danger in overemphasizing the contribution of more remote ancestors. Selection on the basis of pedigree therefore has its most important place in the initial screening of young bulls for progeny testing.

Female selection

Culling

The average herd life in the UK is about four lactations and therefore the intensity of selection for milk production is generally low in most herds. The culling of cows will only contribute to genetic improvement if they are culled for low milk production or for other deficiencies associated with milk production. Unfortunately most culling is carried out for management purposes (Table 2.2), with only 25 per cent of culls accounted for by low yield.

The rate of genetic improvement brought about by the culling of cows thus depends on the reason for and rate of culling, the genetic merit of the replacement heifers and the age of heifers at first calving.

Economic constraints often affect the rate of culling. A high rate of culling to improve the genetic merit of the herd in a particular trait means rearing a greater number of replacements and these may be occupying land which could be used for more productive purposes such as carrying additional milking cows. It has been estimated that only 6 per cent of genetic improvement can be expected from female selection compared with over 30 per cent from the selection of dams of future young sires and over 40 per cent from sires of future young sires. However, although the contribution of culling to genetic improvement is small relative to other selection mechanisms, it has an important economic role in maintaining the current level of milk production in a herd.

Family selection

In pedigree herds family names are normally traced through the dam. For example, the daughters of a foundation cow named Daisy will all have the name Daisy following their herd name, and subsequent granddaughters will also carry the name. Many breeders believe that particular attributes of conformation and temperament are passed through the female line and therefore particular attention is paid to family selection.

Unfortunately half of the genes for a trait are contributed by the sire, and if one considers the diluting effect of the genes of the sire with each successive generation, it is clear that in very few generations the contribution of female (or male) ancestors is minimal. Greater attention should thus be paid to the attributes of near, rather than distant, relatives.

Breeding systems

Crossbreeding

The mating of a purebred from one breed with a purebred of another breed is termed crossbreeding. This system of breeding is used either because of a desire to change the breed, e.g. Ayrshire to Friesian, or to take advantage of the hybrid vigour (heterosis) which results.

The effects of heterosis are most striking in traits with a low heritability such as reproductive performance, and those concerned with vigour. In crossbreeding experiments to examine the effects on milk production, the first crosses between two breeds normally produce 5–10 per cent more milk than the average for the parent stock in the first lactation. The differences for subsequent lactations are smaller.

The major problems associated with crossbreeding are firstly that purebred parents have continually to be found to produce the heterosis and secondly, most farmers have an attachment to one breed.

Inbreeding

The mating of individuals which are more closely related than are the average individuals of the breed, is known as inbreeding. The degree of inbreeding is measured by the inbreeding coefficient F which ranges from 0 to 1 (or 0–100 per cent).

The effects of inbreeding are in many ways opposite to those of heterosis; there is a decline in reproductive performance and vigour. Also there is a reduction in body size and an increased chance that hereditary malformation or lethal genes will appear. Milk and fat yield are reduced through inbreeding by about 0.5 per cent per 1 per cent increase in the inbreeding coefficient, but fat percentage is unaffected. Linebreeding is a system of inbreeding practised by some pedigree breeders in which closely-related individuals are mated in order to maintain desirable characteristics of the line.

Breeding programmes

The first decision that a farmer should make in deciding which breeding programme to follow, is what proportion of his income and profit is derived from milk sales, cull cow sales, calf sales and breeding stock sales, and therefore which traits to select for. If he is a commercial milk producer then the sale of milk fat and milk protein will be his major source of profitability. If he is a pedigree breeder with a reputation for producing cattle of good conformation then the sale of breeding stock will be an important source of income.

The commercial milk producer will therefore make the most genetic progress by selecting and using proven AI sires with high ICCs for milk fat + milk protein yield. The ICC should have a weighting of at least 30 with daughters in at least 10 herds in one calving season. In view of payment penalties on milk fat content, the daughters should have an average value of at least 3.9 per cent. Also the sire should not have deficiencies in the temperament of his daughters, nor in their conformation traits relating to production such as feet, udder attachment and teat placement. Although some female selection should take place for various traits, it is a mistake to have a high replacement rate such as over 25 per cent (the national average), as this requires more replacements to be reared which take up land which could be used for more productive enterprises. Also a high replacement rate results in a greater proportion of heifers and young cows in the herd, with a consequent lower annual milk yield.

Type should only be considered as a primary trait for selection by those pedigree breeders with a large income from the sale of breeding stock. Even then there is a danger of making little genetic progress in milk yield due to the low or negative correlation with most type traits.

Breeding management

3

Reproductive efficiency is a major factor in the profitability of the dairy enterprise, through its effects on the annual milk output of the herd, on the costs of herd depreciation and on the intensity of selection for production traits.

The production of milk is related to the reproductive cycle and hence long intervals between successive calvings lead to reduced annual milk outputs. Low levels of reproductive efficiency give rise to greater numbers of cows failing to conceive and to higher culling rates, which increase the depreciation cost, and reduce the average age of the herd and hence the milk yield per cow. The genetic improvement of a herd is brought about through the replacement of culled cows by heifers of a higher genetic potential, and shorter calving intervals will increase this rate of improvement.

Reproductive efficiency

The objectives in reproductive management are for each dairy cow in the herd to produce a calf annually. In practice, however, only about 85 calvings occur annually per 100 dairy cows intended for breeding (Table 3.1).

The cows which fail to breed include those which do not conceive, and those which conceive but subsequently lose their foetuses and this results in an annual loss of about 8 per cent of potential calvings. The majority of herds have a

Table 3.1 Average levels of reproductive efficiency

	(%)
Cows intended for breeding	100
Calving losses due to failure to breed	8
Calving losses due to prolonged calving intervals	7
Average number of calvings per annum	85

390 days. This gives rise to an annual loss of 7 per cent of the potential calvings.

Artificial insemination

Artificial insemination (AI) is used by progressive dairy farmers as a means of increasing the genetic merit of their herd through the use of semen from progeny-tested bulls of high genetic merit (ICC). If natural service is used the bull is generally unproven and any benefits which might accrue from natural service in enhanced pregnancy rates must be set against the slower rate of genetic progress. It has been estimated that the genetic merit of dairy cattle bred by AI is about 5–7 per cent better than cattle of the same breed bred by natural mating.

Other advantages of AI are that it eliminates the dangers and costs involved in keeping bulls, it offers a wider selection of bulls, and it eliminates the risk of acquiring venereal diseases. The greatest drawback to AI is the need for efficient detection of oestrus.

The pregnancy rate resulting from services by the same bull will be greater when used for natural mating than when used for AI. The reasons for this are that a certain proportion of cows will be wrongly diagnosed as being in oestrus and inseminated, resulting in a reduced pregnancy rate, and some cows will be inseminated too early or too

prolonged calving pattern of up to 9 months, and there is little scope for reducing the number of cows culled for failure to breed.

The culling of problem breeders is beneficial in the long term as it is a method of selecting for higher conception rates, although the rate of genetic progress will be slow because reproductive traits have low heritabilities. In a small number of herds, failure to breed is a major problem and this occurs either where there are infections of the reproductive tract with such organisms as *Brucella, Salmonella* and *Vibrio foetus* or where there is a nutritional imbalance.

In herds with a block calving pattern, where it is intended that all the cows calve within a 10–16 week period, failure to breed within this period can be high (up to 20 per cent) if management is not of a very high standard.

A prolonged calving interval is common for cows in most herds, and in the UK the average herd calving index is over

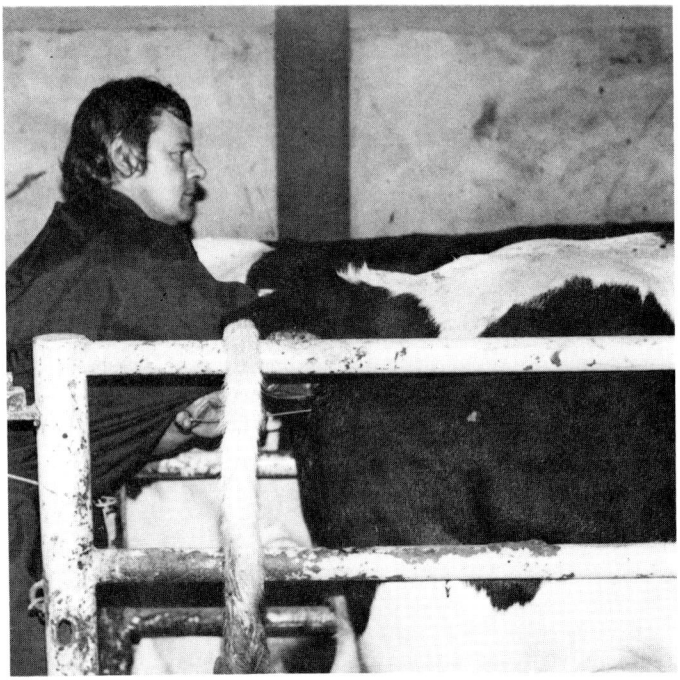

Figure 3.1   The artificial insemination of a dairy cow

late in the oestrous period. Estimates of the difference in pregnancy rates between natural service and AI for bulls of good semen quality range from 2 to 10 percentage units. In practice, however, the choice is often between a home-based bull used for natural service whose semen characteristics have not been tested, and a selected AI bull of high-quality semen characteristics. In such cases there will be similar pregnancy rates for both methods.

In addition to semen quality varying between bulls, the semen quality of the same bull may vary from day to day and from month to month. For this reason it is important that a satisfactory quality control is implemented at the freezing units of the AI stations.

After collection the semen is examined microscopically for the proportion of live spermatozoa. Satisfactory collections then have a diluent added, the amount varying according to the number of sperm present. The semen is put into 0.25 cc straws and then frozen in liquid nitrogen, each straw containing approximately 20 million spermatozoa. Some straws of each batch are thawed and tested for the number of live sperm and for abnormalities. All semen remains in quarantine for 28 days after freezing to ensure that the bull was not incubating any disease at the time of collection. The recovery rate of the semen is further checked by examination of a thawed sample. At each stage of testing, batches of semen may therefore be discarded if found to be of unsatisfactory quality.

Pregnancy rates for artificial insemination are given as 30–60 day non-return rates and average 70–75 per cent. However, many of these supposed pregnant cows are not pregnant, and are not put forward for service again, because they have either been culled, or been re-served with a bull. The ultimate calving rate to first service averages about 55 per cent with a range between herds of 35–75 per cent.

Calving index

The main financial incentive for reducing the calving index from the national average level is that it leads to a greater output of milk per annum, as illustrated in Fig. 3.2. The optimum index for most herds is 340–370 days and although it is tempting to allow high-yielding cows to have an extended calving interval, this is generally not economically beneficial. The relationship between the annual milk yield and the calving interval depends on the shape of the lactation curve of individual cows. For those with a flat curve and high persistency, a short calving interval is less critical for optimum annual milk yield. Hence for heifers there is often little detriment in having a calving interval of up to 13 months, and it is for high yielding cows with high peak yields that the losses from long calving intervals are greatest. It is important, however, that the body condition

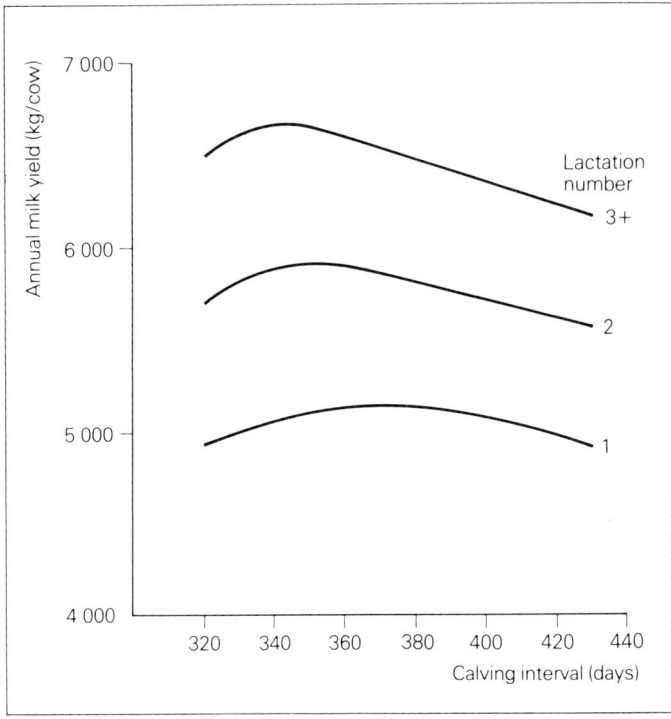

Figure 3.2   The relationship between the calving interval of a cow and its annual milk yield (in first lactation heifers, the highest annual milk yield occurs at a longer calving interval than in older cows)

of the cow is recouped in time for the next calving, otherwise a short calving interval can lead to a reduced yield in the subsequent lactation.

A management advantage of achieving a 12-month calving pattern is that a block calving system can be practised, with all cows in the herd calving in a 10–12 week period. The feeding and breeding management, and the labour organization are thus simplified.

The calving index for any herd can wholly be accounted for by the following three management factors:

1. The time from calving to the average time when it is decided to start serving (this will range in individual herds from under 6 to over 10 weeks post-calving).
2. The oestrus detection rate.
3. The pregnancy rate per service.

The interrelationship of these factors is illustrated in Fig. 3.3. A herd with a calving index greater than 365 days is thus failing in one or more of these management areas.

It is common practice to commence serving at about 63 days *postpartum* and it is clear that this practice makes it more or less impossible to achieve the desired target of 82 days calving to conception interval. Commencing serving earlier than 63 days is the simplest method of reducing the calving interval, and for most situations a 42 day start is advantageous.

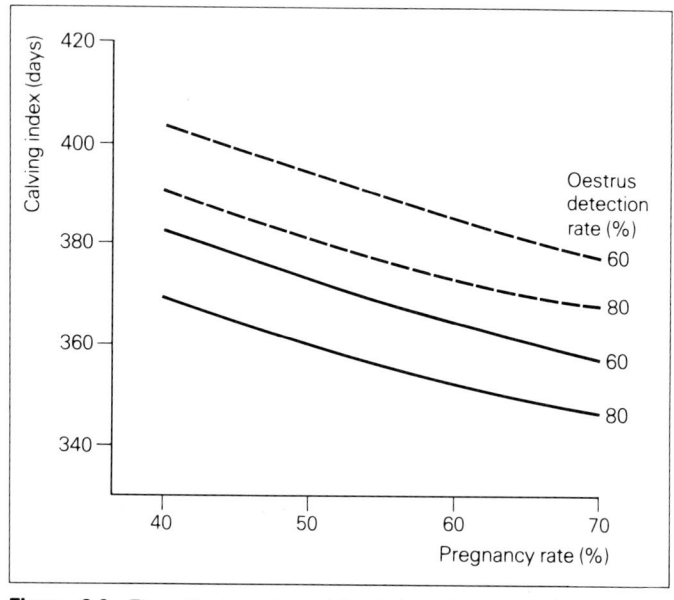

Figure 3.3 The effect on the calving index of commencing serving 42 (——) or 63 (— —) days postcalving, and of the pregnancy rate and oestrus detection rate

Onset of oestrus

In the dairy cow, involution of the uterus may continue for

up to 6 weeks post-calving although ovarian activity normally recommences in the first 3 weeks. This can be seen from an examination of the milk or blood levels of the hormone progesterone, which is produced by the corpus luteum formed in the ovary following ovulation, as illustrated in Fig. 3.4. After one or more short oestrous cycles, the normal cyclical activity is resumed, with ovulation occurring on average every 21 days (normal 18–24 days). Oestrous behaviour is normally observed in only about 25 per cent of cows at the first ovulation but with good management this should increase to over 80 per cent by the third ovulation.

For most cattle, therefore, the first observed oestrus will occur during weeks 3–6 post-calving, but there is a considerable variation between cows and herds. Those cows not seen in oestrus by 6 weeks post-calving should be examined by the veterinary surgeon. The milk yield level of the cow is often thought to be a factor determining the onset of oestrus, but providing the cow is fed to appetite on a balanced diet it is unlikely that the onset of ovarian cycles will be delayed in most high yielding cows by more than 10–14 days compared with cows of average yield levels. Delays in the resumption of normal cyclical activity are more commonly caused by abnormalities of the reproductive tract resulting from difficult calvings (dystokia), retained placenta, endometritis, luteal cysts and follicular cysts. If, however, cows are underfed in early

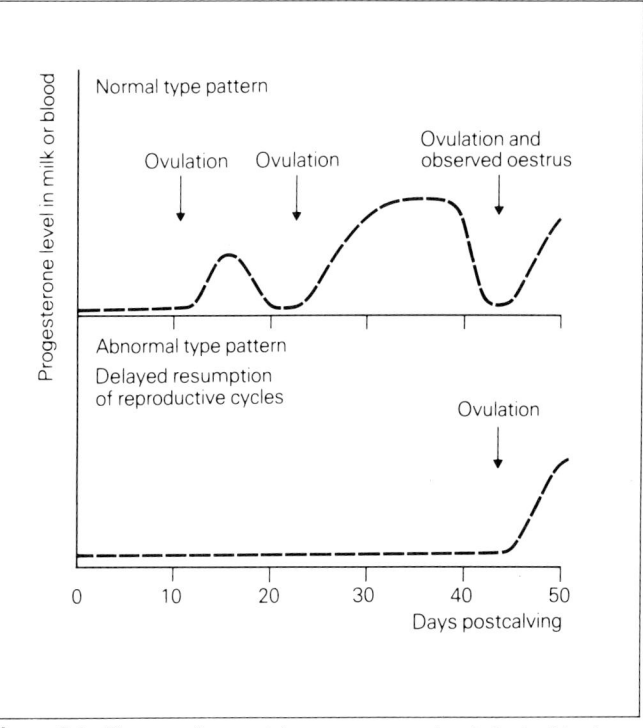

Figure 3.4 Diagram of postcalving progesterone levels in the blood or milk of a cow showing a normal type pattern or an abnormal type pattern (the latter cow could have inactive or cystic ovaries)

lactation and much body condition is lost, the resumption of ovarian activity will be delayed.

Oestrus detection

The oestrus detection rate (ODR) of a herd gives a measure of the efficiency of stockmen in identifying cows in oestrus.

$$\text{Oestrus detection rate (ODR)} = \frac{\text{No. of cows observed in oestrus per 3-week period}}{\text{No. of cows expected to be observed}} \times 100$$

In a number of studies on commercial farms the average ODR has been found to be only 60 per cent. A detailed examination of the reproductive tract of the 40 per cent of cows not observed in oestrus has generally shown that in over 90 per cent of cases the cows were normal and thus there had been a failure to observe oestrus. Only in the remaining 10 per cent was there evidence of abnormalities of the reproductive tract.

Reproductive management where AI is used must therefore put considerable effort into oestrus detection. During service periods, examination of cows at least three times daily is necessary, and one of these times should be in the evening when many cows exhibit clear oestrus symptoms. Unfortunately the duration of oestrous activity may vary from 3 to 30 hours in dairy cows and inevitably some oestrous periods are missed because the cow is in oestrus during the night. However, this level of routine observation should give an ODR of over 75 per cent. If the ODR is less than this, the veterinary surgeon should be brought in to examine cows for any abnormalities of the reproductive tract.

Figure 3.5 Cows showing oestrous activity

A greater manifestation of oestrus is exhibited in cows during spring and summer than during autumn and winter due to the higher temperatures and longer day-lengths. The most important factor, however, is the presence of one or more additional cows in oestrus, which leads to considerably more mounting behaviour thus simplifying detection.

Oestrus detection rates can be enhanced through the presence of a bull sited in the vicinity of the cow yard. Cows in oestrus will generally migrate to the bull pen. Also a vasectomized bull, running with the herd, will identify cows in oestrus and if fitted with a marker under its chin will mark the backs of those cows which are mounted.

Another approach to deal with cows not seen in oestrus is to apply a heat mount detector to the spine of the cow between the hook bones. The pressure exerted by another cow riding on the back of the cow in oestrus induces a colour change in the heat mount detector. A simpler type of heat mount detector is to apply paint to the tail head of cows, which is rubbed off when the cow is ridden. A major problem with these methods is that some false positives will occur, due to non-oestrous cows being ridden by oestrous cows and also by banging or brushing the detectors on obstructions. Heat mount detectors should therefore be used to draw attention to the cows and they should only be inseminated if other symptoms of oestrus are seen.

An alternative approach to oestrus detection is to control oestrus through the use of hormone-type administrations. The two most common types are prostaglandin and progesterone compounds.

Control of oestrus

To overcome the problems of having to observe oestrus if AI is used, two systems of oestrus synchronization have been developed with the objective of inseminating cattle without visual confirmation of oestrus. The two methods involve controlling the length of the luteal phase of the oestrous cycle either by:

(a) Administering progesterone or progestogen for 9–12 days and then withdrawing the treatment, following which the animal normally comes into oestrus and ovulates in about 48 hours;

or

(b) Injecting prostaglandin F 2α or an analogue of this hormone-type compound, which causes the corpus luteum to regress followed by ovulation in 60–84 hours.

The progesterone or progestogen is generally administered as an impregnated intravaginal coil. It is normal with this treatment to initially administer oestrogen to reduce the life of any corpora lutea which could survive beyond the progesterone treatment period of 9–12 days. After withdrawing the coil, the animal is inseminated 48 and 60 hours later, and in some cases only a single AI at about 48 hours is given.

When prostaglandin or its analogues are used two intramuscular injections are normally given 11 days apart and the cows are inseminated at 72 and 96 hours or at 78 hours only, after the second injection. About 50–60 per cent of the cattle will respond to the first injection, although oestrus occurs 12–24 hours later than that following the double injection system, and there is more variation between cattle in the timing of oestrus. However, it is possible to inseminate at observed oestrus following the first injection, and then to give a second prostaglandin injection only to those animals not seen in oestrus, followed by the normal fixed time insemination(s). Prostaglandin F 2α is an abortive agent and therefore should not be administered to cattle which might be pregnant.

In maiden heifers under good management conditions these synchronization techniques have consistently given pregnancy rates of 60–70 per cent, but in lactating dairy cows much poorer results have often been obtained. The explanations for this are mainly connected with the after-effects of the previous pregnancy and the effects of lactation.

In early lactation uterine involution continues for some time and consequently cyclical activity may not have resumed normally when treatment is given. Progesterone coil synchronization treatment may be more beneficial in these cases in stimulating ovarian activity to commence. Conversely, the administration of prostaglandin to dairy cows with recently commenced reproductive cycles can result in the cessation of activity or abnormal cycle length, and this precludes the use of the double-injection technique which depends entirely on the cow being in mid-cycle at the time of the second injection.

For these reasons, and also because of the additional cost, synchronization is not generally used as a blanket treatment for all dairy cows, and is usually restricted to use on a minority of animals which are found by veterinary examination to be cycling but not showing symptoms of oestrus (silent heat or sub-oestrus).

Pregnancy rates

The pregnancy rate (or conception rate) to each service is normally defined as the percentage of cows served which are pregnant at the time of pregnancy diagnosis (6–9 weeks post-service), and is affected both by nutritional and by non-nutritional factors.

A major factor affecting fertility is the energy balance of the cow. This is the intake of energy minus the energy required for maintenance and milk production. If a cow is in severe energy deficit leading to the mobilization of large amounts of body reserves, there will be a delay in the resumption of cyclical activity post-calving or a cessation of cyclical activity already commenced. At more moderate energy deficits of less than about 30 MJ/day (equivalent to about 1 kg of body weight loss daily), the cow may commence reproductive cycles but have less likelihood of conceiving. This type of reduced fertility is normally associated with a lowered blood glucose level.

There appears to be an interaction between body condition and energy balance in their effects on fertility. Cows in good condition can withstand an energy deficit better than those in poor condition (Fig. 3.6). At the other extreme, where cows with a lower milk potential are well fed and in a fat condition, there is also a danger of a reduced level of fertility.

At normal dietary levels protein does not affect pregnancy rates. However at levels above 16 per cent in the dry matter there is some evidence that pregnancy rates are reduced, whilst at levels below 12 per cent there is likely to be a negative effect on voluntary intake, and consequently energy intakes will be reduced, resulting in an adverse effect on milking and reproductive performance.

Mineral deficiencies have been correlated with lower levels of fertility on some farms, but are not often a major cause. Of the major minerals a phosphorus deficiency or an imbalance in the calcium/phosphorus ratio are commonly

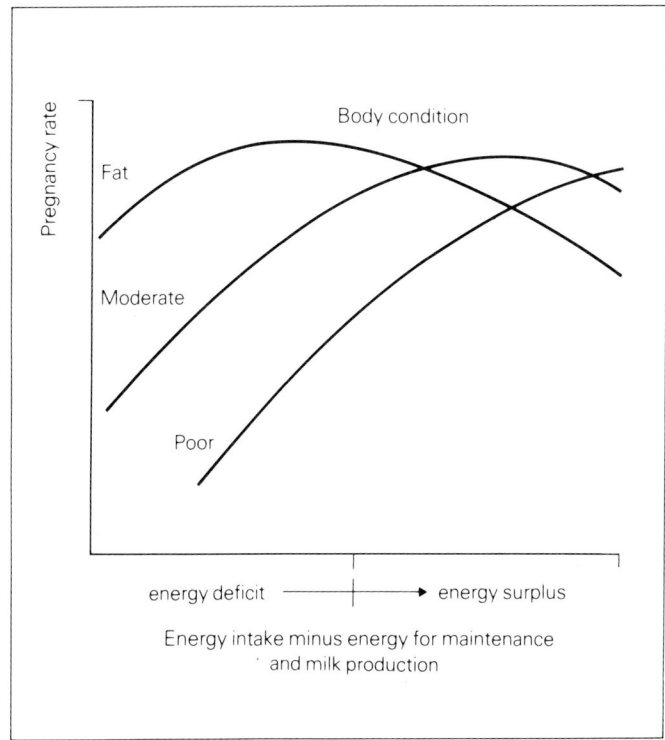

Figure 3.6 The interrelationship between energy balance, body condition and pregnancy rate

cited as causes of reduced fertility, and deficiencies in the trace elements manganese, copper and iodine have been found in herds of low fertility. These deficiencies and imbalances can be detected from a blood analysis, although the critical levels for normal reproductive function have not been fully clarified.

The fertilization rate of ova by spermatozoa in cattle served during oestrus is over 90 per cent although subsequent pregnancy rates only average about 55 per cent. The difference arises from embryo losses which generally occur before implantation is completed at about 30 days. The greatest losses of embryos (about 30 per cent) occur before the fourteenth day post-insemination and these are characterized by a normal return to oestrus at about 21 days after insemination in the same way as fertilization failures. Embryo losses after 14 days account for about 10 per cent of embryos and they result in later returns to oestrus. These later embryo losses tend to be greater for older cattle which suggests a greater degree of implantation failure.

The normal duration of oestrous activity averages about 15 hours and from the end of oestrus to ovulation is about 12 hours. Unfortunately it is difficult to inseminate each cow at the optimum time as there is considerable variation between cows in these durations, and it is very difficult in practice to identify the end of oestrus.

The life of the ovum is quite short, probably only 4–6 hours and it is fertilized in the fallopian tube. The spermatozoa live for approximately 24 hours and they must be present and capable of fertilizing the ovum by 4–6 hours post-ovulation. They reach the oviduct in a matter of minutes following insemination and thawing of the frozen semen, but it appears that they require a short capacitation period before they are capable of fertilizing the ovum.

The optimum time to inseminate for the best pregnancy rates is 12–15 hours before ovulation occurs, which is equivalent to mid to late oestrus as illustrated in Fig. 3.7. Unfortunately as the inseminator only visits the farm once daily it is inevitable that there will be a range in insemination times relative to ovulation and therefore a range in the probabilities of conception.

The storage of frozen semen on the farm and the training of farm staff in insemination techniques should in theory lead to increased pregnancy rates because cows can be served at various times of the day according to the stage of oestrus. However, this advantage has to be balanced against the disadvantage of on-farm inseminators having less experience than full time inseminators. It is likely that do-it-yourself AI will lead to only marginal increases in pregnancy rate on most farms, and its major advantage lies in the saving of the insemination charge.

There is an increase in pregnancy rate with increased calving to service interval up to 10–12 weeks (Fig. 3.8). This is due to the gradual return to normality of the reproductive tract after calving and the improving energy

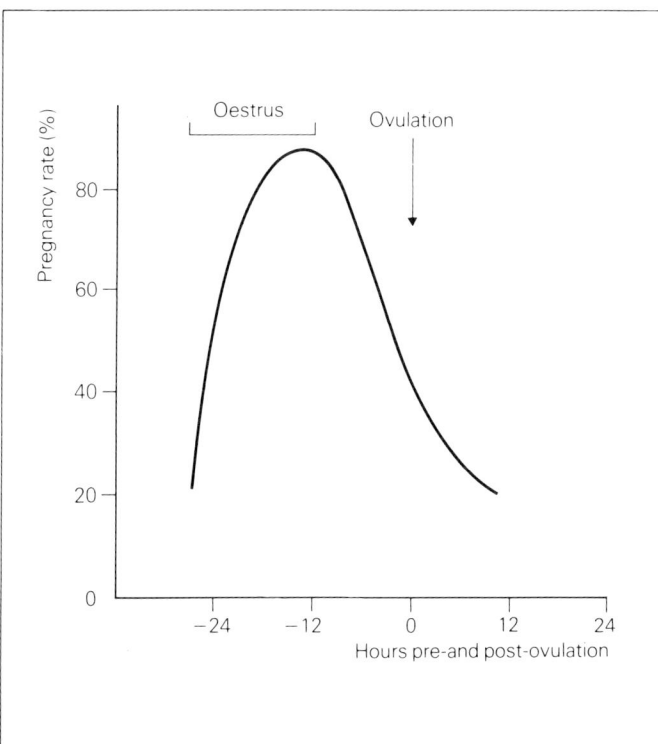

Figure 3.7 The effect on pregnancy rates of the timing of artificial insemination in relation to ovulation

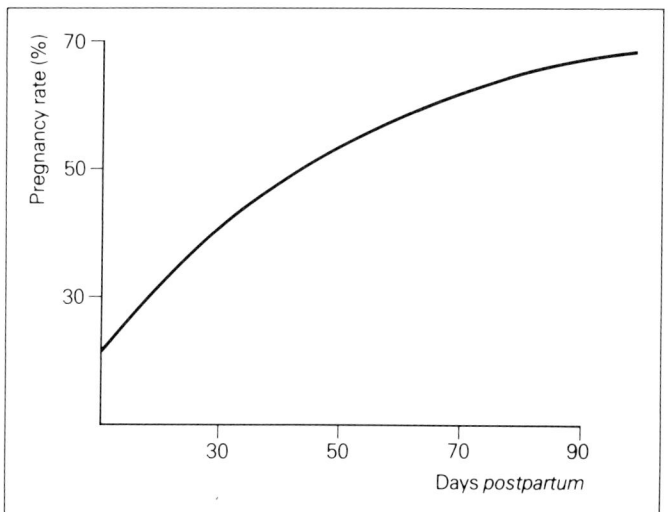

Figure 3.8 The effect on pregnancy rates of the time of insemination after calving

balance of the cow. In spite of this increased pregnancy rate associated with later services it has been calculated that for every day earlier that serving is commenced from 100 down to 42 days post-calving, there is a reduction in the calving interval of 0.7 to 0.9 days. Thus there is little justification for waiting longer than 42 days *postpartum* to commence serving even though pregnancy rates increase.

Pregnancy diagnosis

The main benefit of pregnancy diagnosis (PD) by the veterinary surgeon at 6–9 weeks post-insemination is to have the satisfaction of knowing which cows are pregnant. At least 95 per cent of cows submitted for PD should be pregnant by this time, because those not pregnant (except those with late embryo losses), will have had two or three oestrous periods where observation was possible.

The development of the milk pregnancy test at 22–26 days post-insemination aims to give early recognition of those cows failing to conceive but not observed in oestrus at 21 days. It is based on the difference in the milk progesterone levels between a pregnant cow with a corpus luteum producing high levels of progesterone, and a non-pregnant cow recently in oestrus with low levels of progesterone (Fig. 3.9). The milk samples which are classified positive by the test are on average about 85 per cent correct, and those classified negative are 98–100 per cent correct. The incorrect positives arise from cows having cycles of abnormal length and from embryo losses.

If use is to be made of the test, then it is important that action should be taken on cows with negative pregnancy results. This could involve the application of heat mount detectors or the induction of oestrus with prostaglandin.

Pregnancy

The critical phase in the development of the embryo is in the period from fertilization to implantation in the uterus 30–35 days later. Once implantation is complete the incidence of losses (abortions) for the remainder of pregnancy is normally low, averaging less than 5 per cent. Abortions may result from infections (in particular from brucellosis, vibriosis and leptospirosis) or from nutritional, chemical, hormonal or genetic causes.

A common additional cause of abortion is the artificial insemination of cows during pregnancy. The oestrus symptoms which occasionally are exhibited during pregnancy result from a higher-than-normal oestrogen production. A further cause is where cows are wrongly diagnosed at pregnancy diagnosis as being non-pregnant, and prostaglandin is administered to induce oestrus. This also results in an abortion because it causes the corpus luteum in the ovary to regress, and the progesterone produced by the corpus luteum is essential for the maintenance of pregnancy up to about 200 days of gestation.

Pre-natal development is affected by a large number of factors: hereditary factors from the sire and dam; the size, parity and nutrition of the dam; the position of the foetus within the uterine horn; the presence of another foetus and the size of the placenta. As a consequence the calf at birth may represent from under 5 to over 10 per cent of the

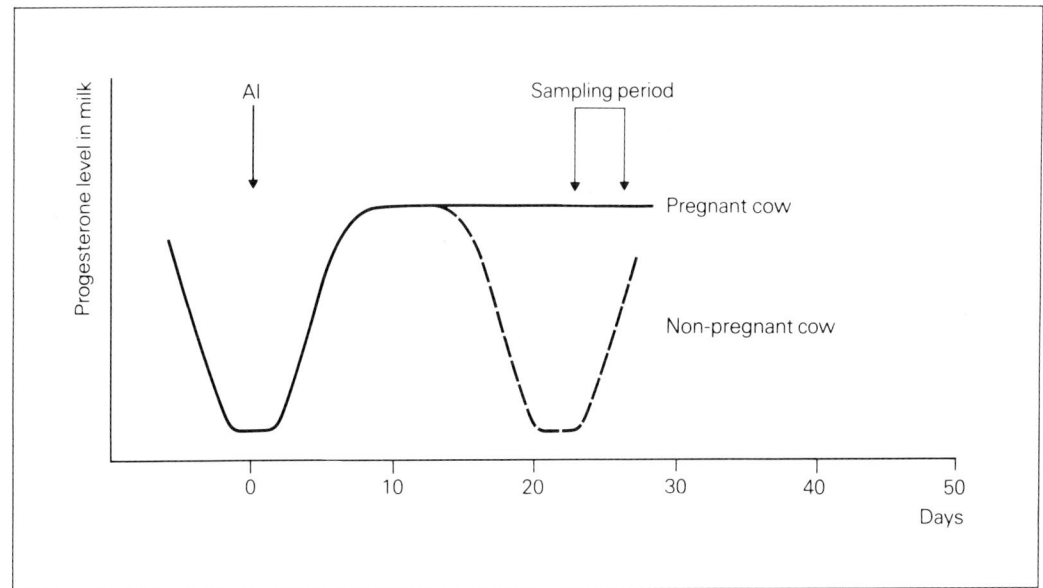

Figure 3.9 Testing for pregnancy from the levels of progesterone in milk

post-calving weight of the dam.

Parturition

During pregnancy the foetus normally lies on its back in the uterus, but prior to parturition it moves into an upright position for the normal anterior presentation. Complications do, however, arise in a minority of pregnancies where there is a posterior (breach) presentation or where the legs or head become twisted backwards.

In the preparatory stage of parturition during which

uterine contractions commence, the cow generally shows signs of restlessness and, outdoors, moves away from the herd. This period can last up to 24 hours but generally is less than 6 hours' duration.

In the second stage expulsion of the foetus occurs. The time between the appearance of the water bag and feet to the expulsion of the foetus is normally less than 3 hours and averages about an hour. This can be prolonged in cases of dystokia resulting from oversized calves or abnormal presentations of the foetus. The foetal membranes are still attached to the calf at birth and continue to supply oxygen to the calf until it can breathe independently. If, however, the second stage is prolonged there is an increasing probability that the calf will suffocate.

In the third stage the foetal membranes are expelled and this normally occurs within 8 hours of birth. If the placenta is retained, the veterinary surgeon will be required to remove it. Retained placentas commonly follow cases of dystokia, premature birth, abortion and multiple pregnancy.

Stillbirths occur in about 5 per cent of pregnancies although in heifers this can rise to over 10 per cent if oversized calves are produced which result in dystokia. The incidence of this type of dystokia can be minimized if a bull of known low dystokia incidence is selected.

Parturition can be induced with pharmacologically active substances such as glucocorticoids and prostaglandins. They

Figure 3.10 A Friesian cow with a Charolais cross calf, two hours after birth

are often used to reduce the gestation length in cattle with suspected oversized calves, e.g. in maiden heifers, or where bulls from exotic breeds have been used. In New Zealand, where the calving pattern is seasonal and very compact, induced parturition is also used on cows conceiving after the desired time in order to bring them into line with the herd calving pattern. Unfortunately this technique does not at present give a predictable response in the time between the administration of the drug and the induced parturition and therefore its use is limited. A further drawback is the high incidence of retained placentas which is often over 50 per cent.

Recording system

It is important if a high level of reproductive efficiency is to be attained to have a satisfactory breeding records system. The essential records for individual cows include the dates of:
1. Calving.
2. Oestrous periods.
3. AIs or natural services.
4. Pregnancy diagnoses.
5. Veterinary treatments.

Various types of system are available. Circular wall charts are useful as indicators of where action is required, but are of less use for historical analyses of reproductive performance. Horizontal wall charts with a line for each cow probably give the best compromise in these two areas of management. The recorded information will alert the herdsman to when the cows are not cycling, when cycles are of abnormal lengths and, historically, it provides the necessary data to identify where any problems lie in reproductive efficiency. The information can also be used to determine when the veterinary surgeon should be called to treat abnormal cows and for pregnancy diagnoses, although in larger herds he may have routine visits for breeding management.

In large herds where much time has to be spent with manual systems there may be justification either for joining a computerized recording scheme — the farmer supplies the weekly records and the organizers of the scheme return a weekly action sheet and provide regular analyses and summaries of the breeding data — or for investing in a microcomputer for the farm, which carries out the same types of analyses as the centralized computer systems.

Principles of feeding

4

Food accounts for about 60 per cent of the total costs of milk production, and therefore the programme of feeding over the lactational cycle is a major determinant of profitability.

The high-yielding cow eats over 5.5 tonnes of dry matter (DM) per annum, of which over half may consist of fibrous foods in the form of grazed herbage and conserved forage. The digestive system of the ruminant is uniquely adapted for the purpose of digesting these food sources which are relatively indigestible to monogastric animals. The remainder of the diet consists of concentrate feeds which are necessary for high daily energy intakes, and which also allow the dairy farmer to carry more cows on the farm commonly termed 'buying acres'.

Food and its analysis

The types of food offered to the dairy cow can loosely be divided into: grazed herbage (grasses, legumes), conserved forage (silages, hays) and concentrates (cereals, crop by-products). To assess the nutrient content of these, a chemical analysis known as the 'Weende Proximate Analysis' is carried out (Fig. 4.1), and the following constituents determined: water, ash, protein, ether extract, crude fibre and nitrogen-free extract. Examples of the proximate analysis are given in Table 4.1.

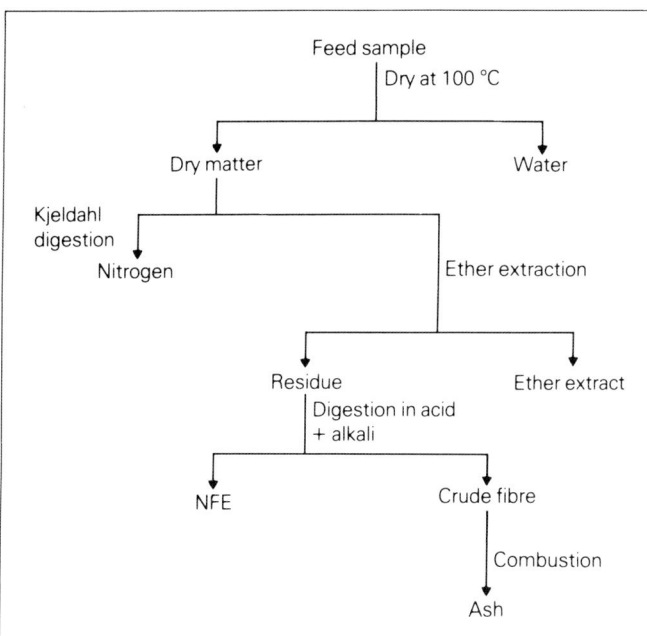

Figure 4.1 The Weende system for the proximate analysis of feeds

Table 4.1 Proximate analysis of some feeds

Food	DM Content g/kg	Analysis of dry matter (g/kg)				
		Crude protein	Ether extract	Crude fibre	NFE	Ash
Grazed pasture	200	225	65	155	465	90
Grass silage	200	170	40	305	390	95
Hay	850	85	16	328	496	74
Barley straw	860	38	21	394	493	53
Barley grain	860	108	17	53	795	26
Soyabean meal	900	503	17	58	360	62

Source: MAFF, DAFS, DANI (1975), *Energy allowances and feeding systems for ruminants*, Technical Bulletin 33, HMSO, London

Water

The water content of feeds ranges from about 10 per cent for dried herbages and certain by-products to over 90 per cent for some root crops. The amount of drinking water required by the cow thus depends on the water content of the diet as well as the environment temperature and the daily milk output. The daily water requirement is generally in the range 30–100 kg.

Ash

The ash, mineral or inorganic matter content is that portion of the food which remains after combustion of the organic matter at 650 °C.

Protein

Proteins contain carbon, hydrogen, oxygen and nitrogen, and in addition may contain sulphur, phosphorus or iron. They are made up of amino acids which are the end-products of protein digestion. The Kjeldahl analysis determines the amount of nitrogen in the food sample, and as proteins usually contain 160g/kg nitrogen, this is multiplied by 6.25 to obtain the crude protein content. The latter includes a group of substances which are not true proteins, these are the non-protein nitrogen compounds such as urea and amines.

Ether extract

This extract contains fats plus other substances soluble in ether. Fats are a very concentrated form of energy containing carbon, hydrogen and oxygen and are also a source of fat-soluble vitamins.

Crude fibre

This represents the less digestible carbohydrate portion of the diet. It includes the majority of the cellulose and lignin of the food. The crude fibre analysis underestimates the fibre content of some foods and is a slow laboratory technique. Other estimates of fibre are now used, the most common being the modified acid detergent fibre method (MAD fibre).

Nitrogen-free extract

This includes the soluble carbohydrates such as starches, sugars, organic acids and some cellulose. It is calculated by difference, subtracting the former components (g/kg) from 1 000.

Digestive system

During the process of digestion, feeds are broken down into small particles and eventually into solution so that absorption of nutrients can take place. This process involves the mechanical breakdown by chewing, microbial action in the reticulo-rumen and to a lesser extent in the large intestine, and enzymatic digestion in the abomasum and small intestine. A diagram of the alimentary tract is shown in Fig. 4.2.

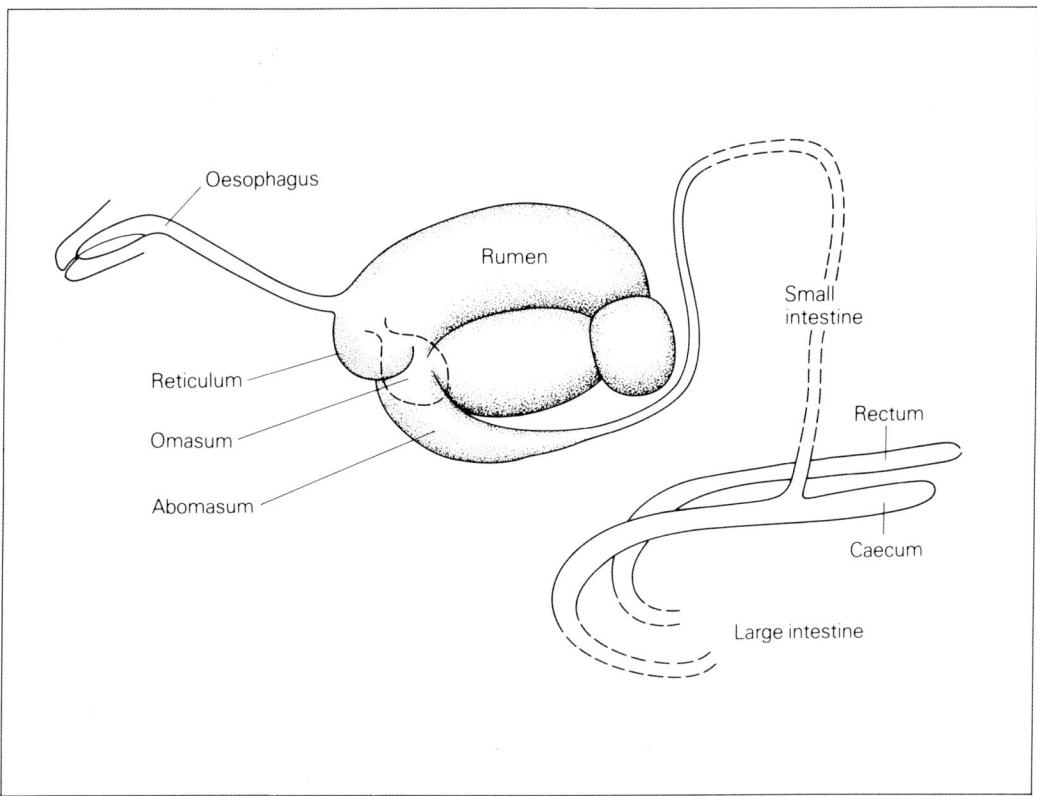

Figure 4.2 A diagram of the alimentary tract of a cow

Mouth

The initial function of the mouth is the grasping (prehension) of food with the aid of a strong rough tongue and eight incisor teeth on the lower jaw in conjunction with the dental pad on the upper jaw. The food is masticated by the molar teeth and mixed with saliva, large quantities of which are secreted (over 100 litres/day). Saliva has a high pH due to the content of bicarbonate/phosphate buffer, and is also the medium for conveying urea to the rumen in the urea recycling process. No enzymes are contained in the saliva of the cow.

Another function of the mouth is chewing the cud (rumination). After a period of storage in the rumen, a bolus of coarse food material is regurgitated to the mouth and masticated for a period of about 1 minute. The bolus is then swallowed and the process repeated. Rumination normally occupies about 8 hours of the day.

Rumen

The rumen is a large organ divided into four compartments with a capacity of about 150 litres. The muscular membrane of the rumen wall is covered in papillae, and this causes a churning motion which mixes the floating coarse particles into the liquid lower half of the rumen.

The main function of the rumen is to act as a large fermentation vat in which micro-organisms digest the carbohydrates of the diet. The bacteria are the most important microbes but they are assisted by protozoa and yeasts. The balance of types of microbes changes with the type of diet, and it takes 1 to 2 weeks for a new population to be established after changing from one diet to another of different type. Gradual changes of diet are therefore important to prevent digestive upsets. The total numbers of microbes vary according to the supply of energy and protein in the diet. The pH is maintained in the range 6.5 to 7.5 by the buffering action of the saliva passing into the rumen.

Carbohydrates such as starch and cellulose are hydrolysed by the microbes into monosaccharides which are then fermented to the organic acids – acetic, propionic, butyric and a small amount of longer-chain acids. The less digestible portions of the diet such as lignin, remain undigested and eventually pass out of the rumen.

The organic acids or volatile fatty acids (VFA) are absorbed through the rumen wall into the bloodstream. The relative proportions of these VFA which are produced varies with the type of diet. High-fibre diets tend to favour acetic acid representing about 60–70 per cent of total VFA, with 16–20 per cent propionic and 7–12 per cent butyric acid. Reducing the fibre content by substituting concentrates leads to a fall in acetic acid to 50 per cent or less, and at the same time to an increase in propionic acid up to as much as 40 per cent. The butyric acid proportion also on occasion increases as acetic acid decreases. These variations in the molar proportions of VFA have important implications in energy metabolism, influencing both milk yield and composition. Diets producing a high proportion of propionic acid tend to favour body fat deposition at the expense of milk yield, and milk of a low fat but high protein content.

The importance of microbial digestion of carbohydrates is indicated from measurements which show that 40–80 per cent of the dry matter intake disappears in the rumen of which 80 per cent is carbohydrate. By contrast, only a limited amount of fats disappear in the rumen. Some of the VFA do, however, arise from microbial action on proteins and other nitrogenous compounds.

The microbes as they multiply synthesize proteins for their own bodies from the non-protein nitrogenous compounds in the rumen. This microbial protein is subsequently digested when it passes into the abomasum and small intestine. A further function of the microbes is to synthesize B complex vitamins.

Large amounts of methane and carbon dioxide and smaller amounts of other gases are produced as a result of the fermentation process. This gas is mainly eliminated from the rumen by the regular reflex action of belching.

Reticulum

The reticulum is a sac lying anterior to the rumen and not completely separated from it, so that digesta can pass freely between the two organs. The internal surface has a honeycomb appearance and contains no secretory cells. It is directly concerned with the passage of the bolus of food up the oesophagus during rumination, and of digesta from the rumen to the omasum.

Omasum

Like the reticulum and rumen, the omasum has no secretions. It is round in shape with a capacity of 15–20 litres, and has muscular leaves on its internal surface. A large amount of water and some organic acids are removed from the digesta during passage.

Abomasum

This is the true stomach and has a similar capacity to the omasum. The normal gastric juices are secreted: hydrochloric acid which changes the medium for the digesta to acid; pepsin, an enzyme which acts on proteins breaking them down to peptides; and rennin, which in the young calf forms clots of the ingested milk.

Intestines

The small intestine is about 40 m in length and its surface is covered with finger-like villi. It is the principal site of absorption for amino acids, lipids, vitamins and minerals. The secretions of the small intestine and pancreas contain enzymes for the further digestion of carbohydrates, fats and proteins, and this takes place in the alkaline medium of the bile.

In the large intestine which includes the caecum, colon and rectum, further bacterial degradation of undigested food residues continues. There are no enzymatic secretions but much water is absorbed, and it is an important site for the absorption of sodium, chloride, potassium, phosphorus and magnesium. The undigested residue is eventually passed out at the anus as faeces.

Nutrient requirements

The measurement of the requirement for different nutrients at different performance levels allows suitable diets to be formulated to meet these requirements. The requirement may be defined as the amount of nutrient which must be supplied in the diet to meet the animal's needs for maintenance of the body, and for production (lactation, reproduction and growth).

Energy

The term 'energy' describes the property of matter to 'do work'. All forms of energy can be converted into heat, and the measure of heat production of a diet is known as its gross energy (GE) content. Carbohydrates have a GE value of about 17.5 MJ/kg dry matter, fat has about 2.5 times and protein 1.5 times as much energy. As carbohydrates are the major portion of cows' diets, GE values average about 18 MJ/kg dry matter.

Of the GE eaten, only a relatively small proportion subsequently becomes available to satisfy energy requirements. Losses of energy occur as the food passes through the body during the processes of digestion and metabolism, in the faeces, urine, as gas and as heat (Fig. 4.3).

The loss of energy in the faeces varies considerably for different feeds, ranging from about 60 per cent of the GE for straws to only 15 per cent in some concentrated cereal-based feeds. The urinary energy loss varies from 2 to 8 per cent of GE, the higher losses occurring where excess protein is fed resulting in the excretion of nitrogen. The gaseous energy loss, mainly as methane, accounts for 6–10 per cent, and the heat loss 5–8 per cent of the GE. An illustration of these energy losses is given in Table 4.2.

The main variable is the faecal energy loss, and as urinary and gaseous energy losses vary very little, the metabolizable energy (ME) of a food can be calculated with some accuracy from the digestible energy (DE) using a factor of 0.81.

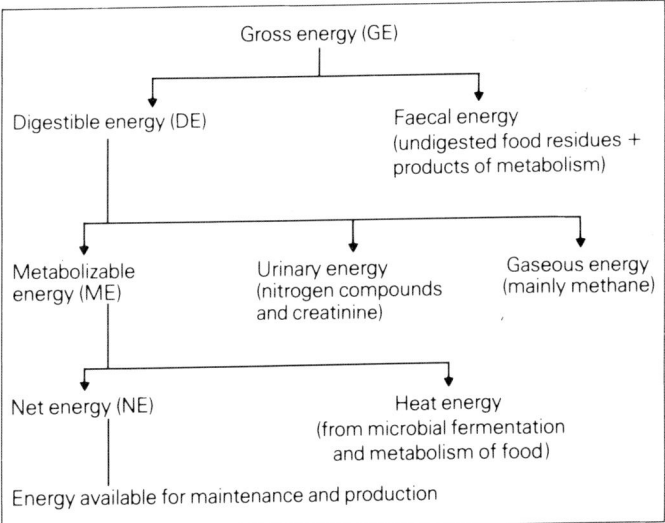

Figure 4.3 The losses of energy in the faeces, urine, as gas and as heat during the digestion and metabolism of a feed

Table 4.2 Example of an energy balance

A cow on energy balance eats 17.5 kg DM of a diet of 18.5 MJ/kg DM GE.

GE intake	324 MJ	
Faecal energy loss	= 81 MJ	
Urinary energy loss	= 18 MJ	
Methane energy loss	= 26 MJ	
DE intake	= 324 – 81	= 243 MJ
ME intake	= 324 – (81 + 18 + 26)	= 199 MJ
ME concentration	= 199/17.5	= 11.4 MJ/kg DM

Table 4.3 Efficiency with which metabolizable energy is utilized for various body processes

ME for maintenance (km)	= 0.72	
ME for milk production (kl)	= 0.62	
ME for gain (lactation) (kg)	= 0.62	
ME for gain (dry period) (kg)	= 0.3–0.6	(ME concentration 7–14 MJ/kg DM)
ME for body tissue mobilization	= 0.82	

Source: MAFF, DAFS, DANI (1975), *Energy allowances and feeding systems for ruminants*, Technical Bulletin 33, HMSO, London

The energy requirements of dairy cattle are thus normally defined in terms of DE, ME or net energy (NE), or in units such as Total Digestible Nutrients, Scandinavian Feed Units, or Starch Equivalent which relate closely to one or other of these energy measurements. In the UK a system based on ME has been developed. An illustration of ME requirements for dairy cattle is given in Fig. 4.4.

The ME of the diet is converted with varying degrees of efficiency for different body processes (Table 4.3). For cows in mid and late lactation laying down body fat, this is

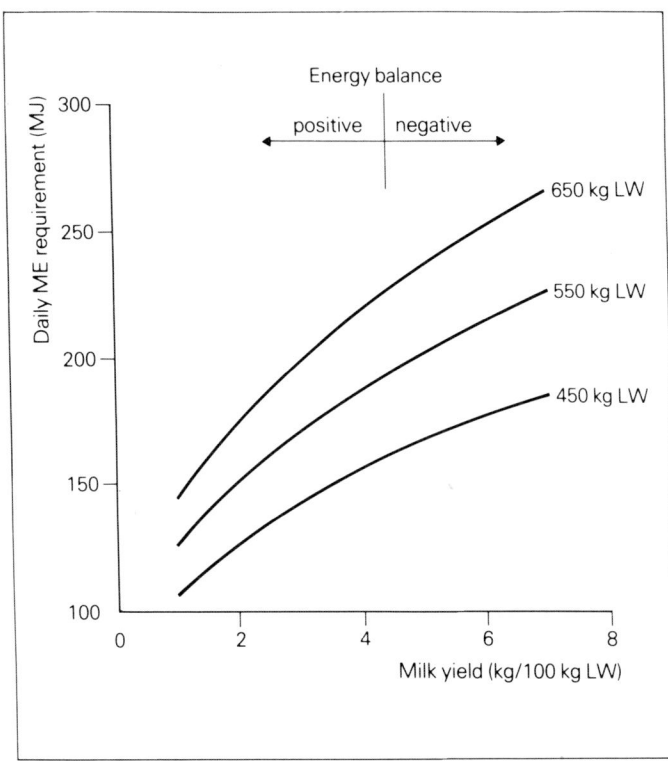

Figure 4.4   The daily metabolizable energy requirements of dairy cattle

carried out with an efficiency equal to that of milk production, but fat deposition in the dry period has a lower efficiency of conversion.

The ME values of foods can be estimated using sheep or cattle on a metabolism trial to measure GE intake and excretion losses of energy in the faeces, urine and as gas. However, to simplify and speed up the assessment, relationships have been determined between ME and various chemical constituents of the feed as follows:

$$\text{ME (MJ/kg DM)} = 14.3 + 0.017\,\text{CP} - 0.019\,\text{MADF}$$

where CP is the crude protein in the dry matter (g/kg)

MADF is the modified acid detergent fibre in the dry matter (g/kg).

An example of ration formulation using the ME system is given in Table 4.4. The ME requirement for maintenance, milk production and liveweight change is first calculated. The contribution to the requirements of the forage intake is then determined, the remainder having to be made up by concentrates. In practice the forage may be fed *ad libitum* and therefore an estimate of forage intake has to be made. Also the cows within a group will have a range of liveweights and yields, and will probably be individually fed on different amounts of concentrates, which in turn will affect individual forage intakes. This type of formulation, because of the assumptions and possible errors involved in the calculation, should only be used for planning purposes.

Table 4.4 A simple example of ration formulation

Ration required for a group of Friesian cows, liveweight 590 kg, milk yield 20 kg/day of 4.0% fat and 8.8% SNF, to give a growth rate of 0.5 kg/day.
Feeds available:
  30 kg/cow of silage (25% DM and 10 ME)
  concentrates (85% DM and 12 ME)

		MJ of ME		
Maintenance requirement	=	62		
Milk (5.27 MJ/kg)		105		
LW gain (34 MJ/kg)		17		
Total ME requirement		184		
Silage (7.5 kg DM × 10 ME)		75		
ME to be supplied by concentrate		109		
Concentrate DM required per cow		= 109/12	=	9.1 kg
Concentrate (fresh weight)		= 9.1/0.85	=	10.7 kg/day

Feeding management should also involve responding on a day to day basis to milk output changes in relation to feed input, in an attempt to optimize the margin over feed costs.

Protein

Milk has a high protein content, accounting for almost 30 per cent of the total solids. The cow also requires protein for maintenance purposes, for growth, for the replacement of tissue protein mobilized in early lactation and for pregnancy.

Protein requirements are often expressed as digestible crude protein (DCP). For a 600-kg cow, maintenance requirements are 350 g/day of DCP plus 55 g/day/kg milk.

Unfortunately digestion of protein in the ruminant is a complex process (Fig. 4.5). The diet contains both true protein and non-protein nitrogen (NPN). All of the NPN and part of the protein of the diet is degraded in the rumen by the micro-organisms to form ammonia, the remaining undegraded protein passing into the abomasum and intestines where it is digested by normal enzymatic action. The bacteria in the rumen use ammonia for growth and this microbial protein continually passes into the abomasum and small intestine where it is digested in the same way as undegraded protein. Any ammonia which is produced in the rumen which is in excess of that required by the microbes is absorbed into the bloodstream, transported to the liver, and converted to urea which is excreted in the urine or recycled to the rumen via the saliva.

The amount of amino acids supplied to the tissues is thus determined by:
1. The amount of protein eaten.
2. The relative proportions of true protein and non-protein nitrogen (NPN)
3. The degradability of the true protein by the rumen microbes.

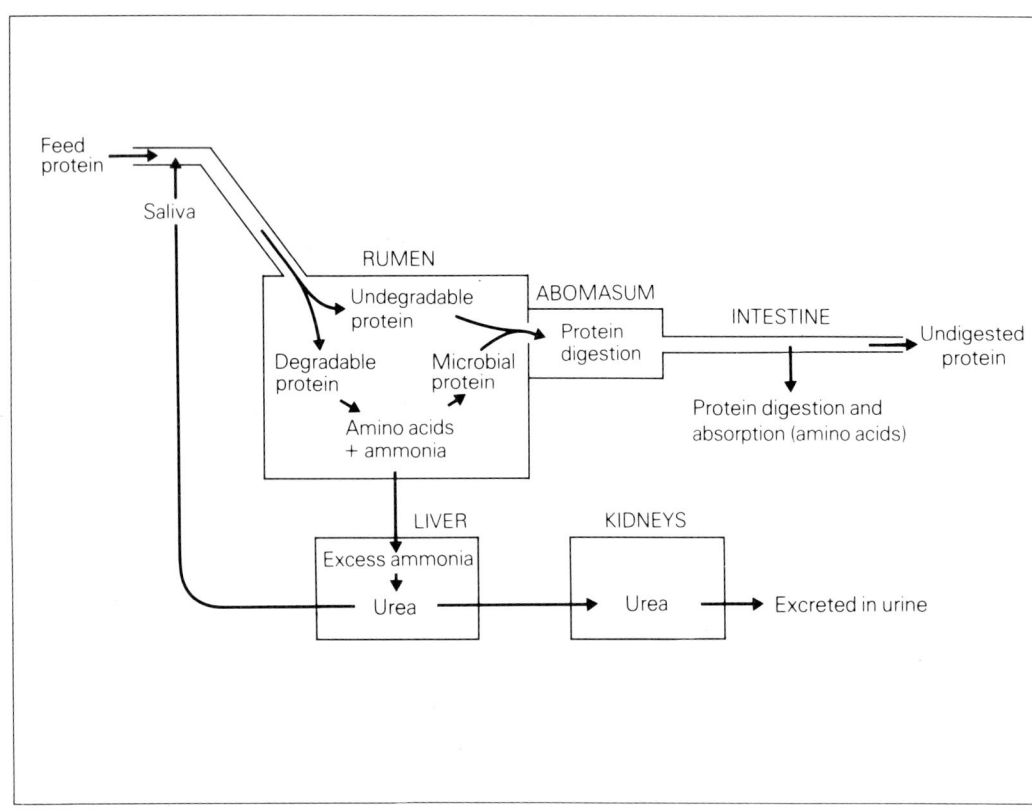

Figure 4.5 Diagram of protein digestion in the dairy cow

4. The requirement of the rumen microbes for ammonia – this is determined by the supply of energy from the diet.
5. The digestibility and quality of the microbial protein.
6. The proportion of indigestible protein in the diet.

Microbial protein production therefore depends on the availability of an adequate energy supply, and rumen degradable protein (RDP) supply from the diet. When protein requirements are high, as in early lactation, microbial protein production is often insufficient to supply the cow's needs and milk production is depressed. In this situation an adequate supply of undegradable protein (UDP) in the diet is required. The choice of protein source in the diet thus becomes very important.

Degradability values range from 100 per cent for NPN sources such as urea to about 30 per cent for fish meal and some heat-treated protein concentrates. The protein of grass silage is highly degradable due to its high NPN content, at about 80 per cent.

Due to the better understanding of protein digestion, the DCP concept of evaluating feeds is becoming outmoded. The quantification of the requirements for RDP and UDP, and the development of simple laboratory procedures satisfactorily to determine protein degradabilities should lead to an improved assessment of protein requirements, and a more accurate rationing procedure.

Minerals

Minerals are required for the basic functions of the body. The minerals which are essential in the diet of the cow are shown in Table 4.5.

The dietary requirements are often difficult to define because interrelationships exist among minerals, and between minerals and organic constituents, which affect their utilization. The amount fed also affects utilization, an excess of mineral may lead to a reduction in the absorption from the gut and to an increase if insufficient is fed.

Mineral requirements are of increasing importance because as pastures are fertilized more highly and greater amounts of herbage are utilized per hectare, imbalances and deficiencies of minerals become an increasing problem. A guideline to recommended levels is given in Table 4.6, and the mineral content of some feeds in Table 4.7.

Table 4.5 Essential minerals in the diet of the dairy cow

Major minerals	Important trace minerals	Other minerals
Calcium	Copper	Chromium
Phosphorus	Cobalt	Fluorine
Magnesium	Iodine	Nickel
Sodium	Iron	Silicon
Chlorine	Manganese	Sulphur
Potassium	Molybdenum	Tin
	Selenium	Vanadium
	Zinc	

Table 4.6   Mineral requirements* of lactating cows (milk yield 20 kg/day)

Major minerals (g/kg DM)		Trace minerals (mg/kg DM)	
Calcium	3.4	Copper	8–11
Phosphorus	3.1	Cobalt	0.11
Magnesium	1.8	Iodine	0.5
Sodium	1.2	Iron	40
		Manganese	20–25
		Zinc	30
		Selenium	0.03–0.05

* *Source: The Nutrient Requirements of Ruminant Livestock*, Commonwealth Agricultural Bureaux, 1980

Table 4.7   Mineral content of some feeds

	In the DM							
	(g/kg)				(mg/kg)			
	Ca	P	Mg	Na	Cu	Co	Mn	Zn
Grazed pasture	5.0	3.0	1.7	1.9	10	0.10	90	25
Grass silage	4.5	2.8	1.3	3.7	8	0.10	90	33
Hay	4.0	2.2	1.2	1.8	5	0.07	100	21
Barley straw	2.7	0.9	0.7	1.1	3	0.04	40	13
Barley grain	0.5	4.0	1.3	0.2	5	0.04	20	34
Soyabean meal	3.5	6.8	3.0	3.8	25	0.20	36	75

Calcium and phosphorus: These two minerals account for 70 per cent of the ash content of the body and their metabolism is closely interrelated. The cow is able to store and mobilize up to 10 per cent of its calcium reserves and so defining precise requirements at any particular time is difficult. It is generally considered that for cows of about 600 kg, feeding levels of about 15 g/day are required for maintenance plus 1.7 g/kg milk produced. Due to the cow's ability to store calcium, deficiencies rarely occur except at parturition when the requirements approximately double with the commencement of milk secretion and removal. If calcium intake and mobilization are inadequate to meet this increase milk fever (parturient paresis) occurs.

Phosphorus is also a major constituent of bone and milk. Recommended minimum feeding levels are about 13 g/day for maintenance plus 1.5 g/kg milk produced. It is generally considered that the ratio of calcium : phosphorus in the diet should be between 1 : 1 and 2 : 1.

Magnesium.   The majority of the magnesium content of the body is stored in the bones and is not easily mobilized. Hence if deficiencies occur as in grazed pasture in spring and autumn, grass staggers (hypomagnesaemia) quickly results. Feeding levels should be about 11 g/day for maintenance plus 0.7 g/kg milk.

Sodium: The dairy cow requires relatively large amounts of sodium, although deficiencies rarely occur. The diet should supply about 5 g/day for maintenance plus 0.65 g/kg milk.

Trace minerals. These are required in relatively small amounts (Table 4.6). In certain areas of the country, however, deficiencies of these trace minerals are increasing, notably in copper, cobalt, manganese and selenium. These tend to occur where farm-produced forage, grazed herbage and/or cereal grain make up a high proportion of the diet. Purchased compound concentrates tend to include supplementary trace minerals which offset these problems.

Vitamins

These are organic compounds which have an important role in dairy cow nutrition. Unlike monogastric animals, ruminants have the ability to synthesize the B-complex and K vitamins in the rumen. Also vitamin C appears to be synthesized in sufficient amounts in body tissue to satisfy requirements. Vitamins A, D and E have to be supplied by the diet.

The precursor of vitamin A is carotene and this is abundant in green feeds such as grazed pasture and grass silage, but concentrate feeds and root crops have negligible amounts. The minimum daily requirements for a lactating cow are about 150 mg/kg liveweight of *b*-carotene.

In contrast green feeds have only low levels of vitamin D. Field-cured hay is a good source of the vitamin. During the summer, sunlight acts on chemicals in the body to produce vitamin D in adequate amounts, but in winter supplementary vitamin D should be added to the concentrate ration. The daily requirement is about 0.25 mg/kg liveweight of vitamin D.

Grass and grass silage are good sources of vitamin E for dairy cows, and smaller quantities are found in cereals and other concentrate feeds. The daily requirement for vitamin E depends on the amount of selenium in the diet, but the requirement can readily be met from normal diets without resort to additional supplementation.

Food intake

One of the main limitations to the productivity of the dairy cow is its voluntary food intake, and this is particularly so in early lactation. The amount of food eaten by a cow on a particular day depends both on factors associated with the animal and on the type of food.

Animal factors

The food intake varies according to the stage of lactation (Fig. 4.6). At parturition, intake is reduced due to the

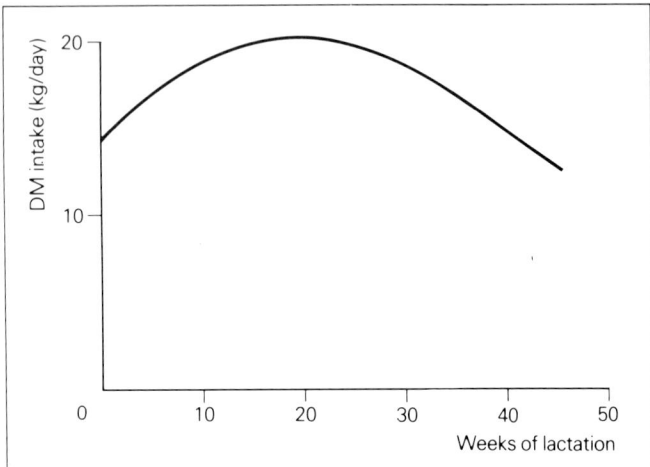

Figure 4.6 The voluntary intake during lactation of a high-yielding cow weighing 600 kg at calving

trauma of calving, but it increases to a peak at 4 to 16 weeks depending on a variety of factors. Peak intake tends to be attained earlier in thinner cows and on diets of higher energy concentration (ME per kg DM). Thereafter intake declines and this is accelerated from the fifth month of pregnancy. Intake also falls off after drying off.

Food intake increases with both body size and milk yield by about 2 kg DM/100 kg liveweight and by 0.2 kg DM/kg milk. An important factor which modifies these relationships is the fatness of the cow. Thin cows eat more than fat cows of the same size.

Food factors

The energy concentration of the diet has an important effect on feed intake. For diets of moderate to low quality such as forage diets, intake is controlled by the physical fill of the rumen and therefore by the rate of flow of digesta from it. As the ME concentration (or the digestibility) is increased the rate of breakdown of fibre by the microbial population increases thus allowing a greater intake of food (Fig. 4.7). Eventually a point is reached where this ceases to be the limiting factor to intake and a metabolic control by the cow comes into effect. The ME concentration at which this takes place will depend on the fatness of the cow, her milk yield and the type of diet. Further increases in the ME concentration of the diet will lead to a reduction in DM intake, and a levelling of ME intake. In general, intakes are greater at a particular ME concentration for diets with a high proportion of concentrates.

When a forage is offered *ad libitum*, supplementation with concentrates leads to a reduction in forage intake. The size of this substitution varies according to the amount of concentrates fed and the quality of the forage – the greater

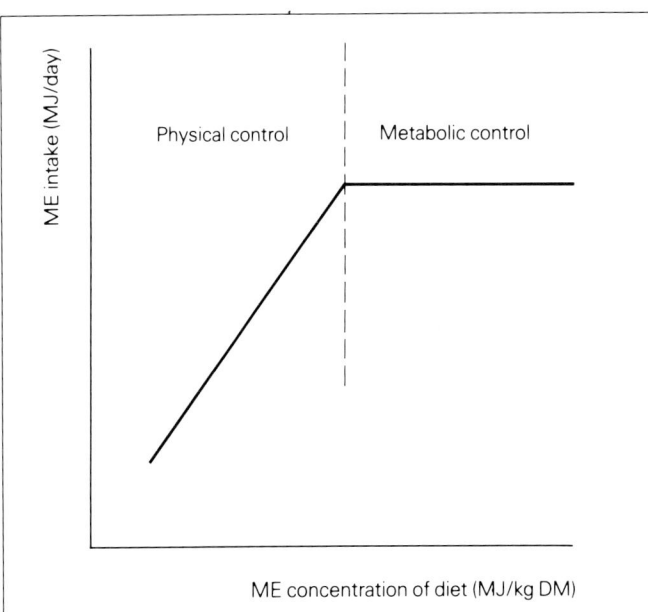

Figure 4.7 The relationship between the metabolizable energy concentration of the diet and the daily intake of metabolizable energy

the digestibility the greater the substitution, ranging from about 10 per cent for low-quality to 90 per cent for high-quality forages.

The intake of diets with a low protein level may be limited, as inadequate nitrogen is available for the microbes of the rumen to grow. An addition of an NPN source such as urea will increase the intake by dairy cows of diets with crude protein contents of less than about 13 per cent in the dry matter if energy is not limiting. It is unusual, however, for dairy cows to be offered diets with such a low protein content.

Response to energy input

When supplementary energy in the form of concentrates is added to the diet of cows offered forage *ad libitum*, a reduction in forage intake occurs but total energy intake is increased. This additional energy is partitioned to additional milk production, to fat deposition, to growth (in the case of young cattle) or to the foetus (in pregnant cattle).

The partition of energy is under the control of the endocrine system. For cows of high genetic potential and for those in early lactation, partition is predominantly to milk, but for cows of low genetic potential and for those in mid to late lactation an increasing proportion of the net energy is partitioned to body fat.

Late lactation and the dry period

This period is of great importance, as the cow partitions an

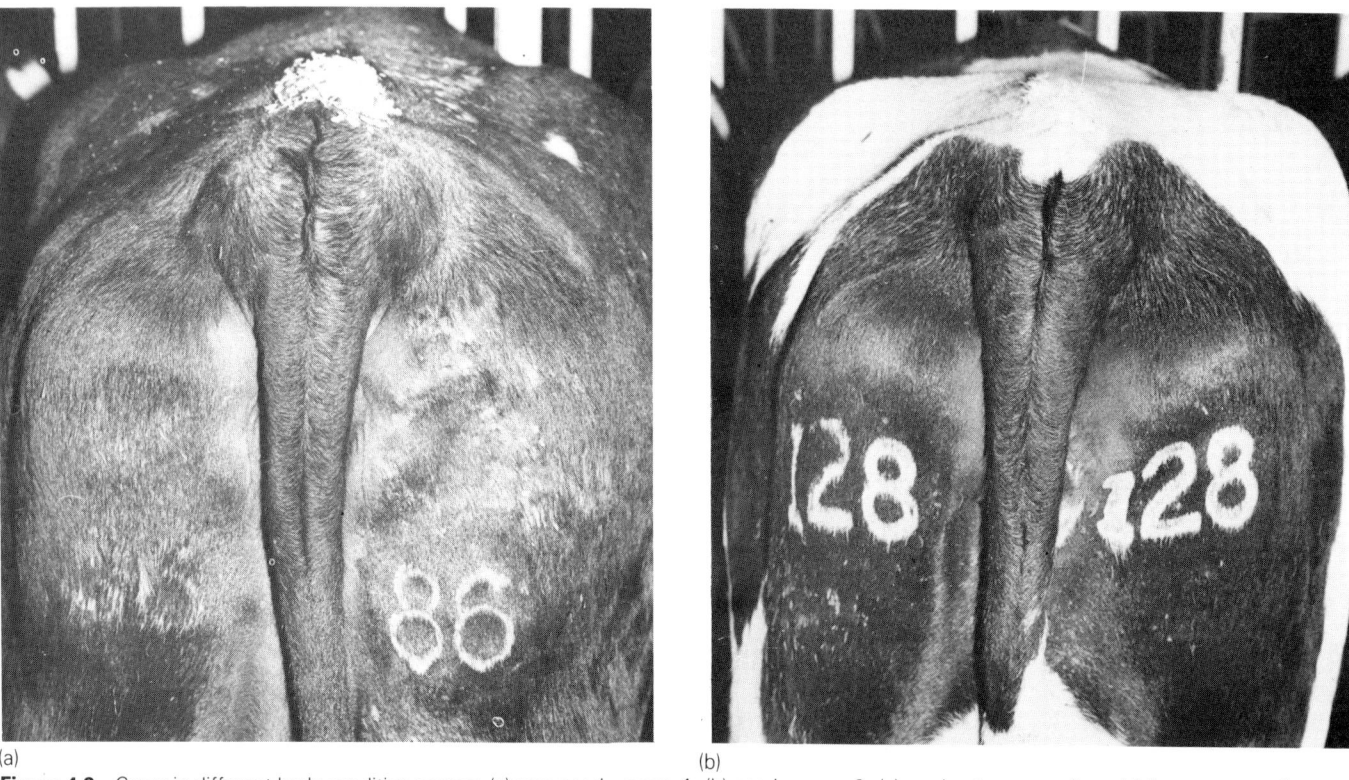

(a)

(b)

Figure 4.8 Cows in different body condition scores: (a) very good – score 4; (b) good – score 3; (c) moderate – score 2; and (d) poor – score 1.

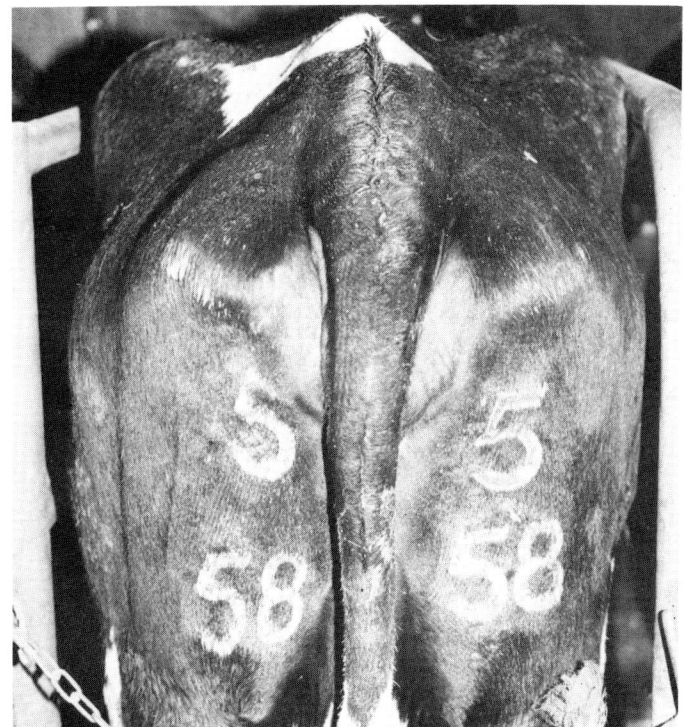

(c)

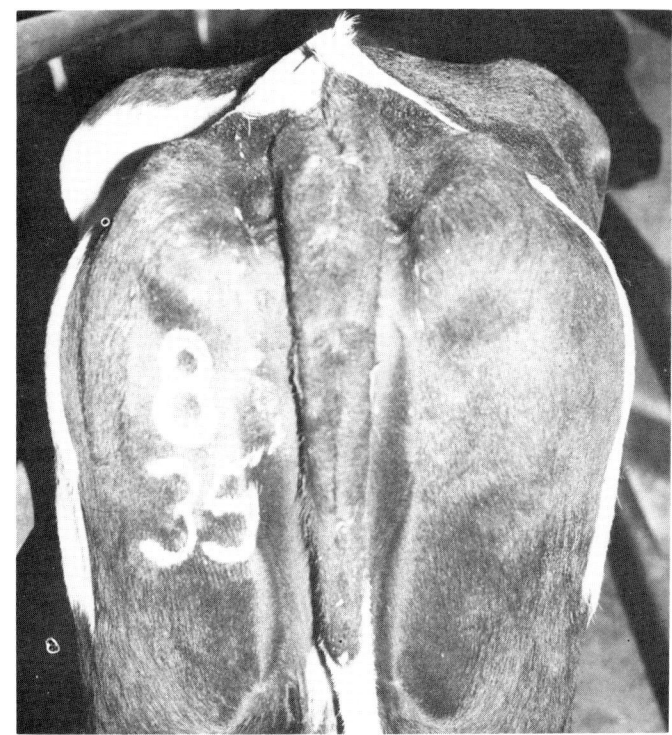

(d)

increasing proportion of its energy intake to the growth of the foetus, to laying down body fat, and after drying off to the generation of secretory tissue in the udder. Underfeeding at this time can have adverse effects on subsequent lactation performance.

A cow of 600 kg liveweight has an ME requirement for pregnancy of about 20 MJ/day during the last month before calving, in addition to its normal maintenance requirement. At DM intakes of below about 8 kg/day, therefore, the cow will lose body condition (fat stores). For heifers and, to some extent, second calving cows, the energy intake in the dry period is even more important than in the mature cow as a proportion is partitioned for growth. A similar minimum DM intake to that of the mature cow is thus required.

Body condition at calving

The body condition at calving *per se* rather than the actual feeding levels in the dry period appears to be the important factor affecting the milk yield in early lactation.

Body condition can be estimated from the amount of fat cover around the tail-head and forward to the loin area (Fig. 4.8). The scale of condition scoring runs from score 0 (emaciated) to score 5 (very fat). Most dairy cows are found in the 1.5 to 3.5 range of score and average targets for calving should be score 3.0–3.5 for British Friesians and 2.5–3.0 for the more extreme dairy breeds.

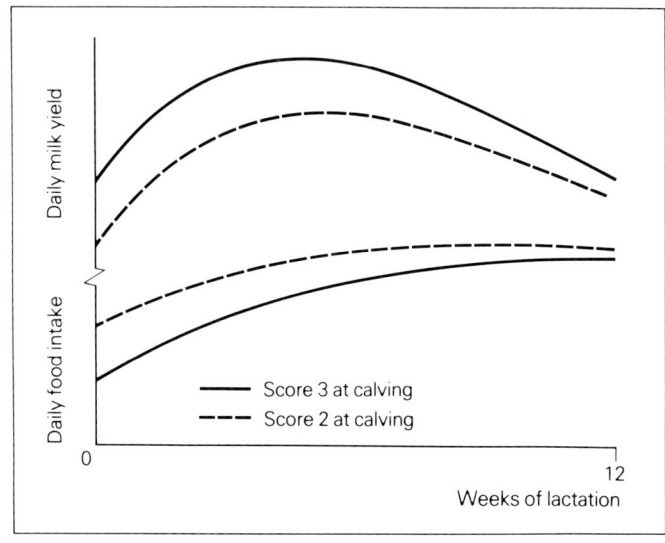

Figure 4.9 The effect of body condition score at calving on food intake and milk yield in early lactation

For heifers and young cows the effect of body condition at calving on food intake and milk yield in early lactation can be substantial (Fig. 4.9). Fatter animals eat less but their peak yield tends to be greater due to the mobilization of body fat. The protein content of the diet is important as it has to balance the additional energy contributed from body fat. For cows which have attained mature size, body condition is not quite so important because their voluntary intake is large enough to support high yields. However, a minimum score of 2.5 at calving should be aimed for.

Stages of lactation

During lactation the objectives of feeding management for high lactation yields are to achieve high peak yields, to maintain this peak as long as possible, and to have a high rate of persistency.

The response in milk yield to an increase in energy intake shows a diminishing returns effect: at greater increments, a greater proportion is converted to body fat. In early lactation the cow may be in energy deficit for some weeks. The factors controlling intake during this period are not clearly understood. If the limitation is the physical capacity of the rumen and the rate of breakdown of fibrous material, then intake can be increased by increasing the concentrate level – this is known as lead feeding. If, however, the control is metabolic then lead feeding can only increase energy intake up to the point where metabolic control operates.

A further problem of feeding high levels of concentrates is that it leads to an increase in the proportion of propionic acid in the rumen at the expense of acetic acid, and this can shift the partitioning of energy towards fattening and away from milk production, the result being reduced lactation yields and fat cows. This is likely to occur when the ME concentration of the diet is greater than about 11.8.

In mid and late lactation a greater proportion of dietary energy is converted to body reserves. It is worth while to recoup reserves then because the efficiency of conversion of ME to body tissue is higher during lactation than in the dry period (Table 4.3 p. 49), and for autumn calving cows this can be achieved on the cheapest food source – grazed grass.

Milk composition

Milk is paid for according to its milk constituents, and feeding has important effects, particularly on milk fat and protein contents.

Milk fat

The average milk fat content varies for different breeds ranging from 3.8 per cent for Friesian to 5.1 per cent for Jersey cattle. Milk fat is made up of fatty acids derived from two sources. Those containing 4–10 carbon atoms are synthesized in the udder from acetate and B-hydroxybutyrate which are mainly derived from acetic and butyric acids produced in the rumen. Those fatty acids in the milk containing 18 or more carbon atoms are directly transferred from the blood triglycerides within the udder. Those fatty acids with intermediate chain lengths derive from both sources.

The milk fat content is affected both by the level of feeding and by the ME concentration of the diet; increases in either or both of these factors leading to a reduction in milk fat content. Those diets which reduce milk fat tend to induce fast rates of fermentation in the rumen, producing an increased proportion of propionic relative to acetic and butyric acids. Adequate long fibre in the diet is necessary therefore to maintain milk fat content, and a minimum of 15 per cent crude fibre in the total dry matter is recommended.

The adverse effects of low milk fat content can be overcome to some extent by feeding the concentrate portion of the diet in smaller and more frequent feeds. However, increasing the frequency of feeding where normal fat levels are being achieved will have no beneficial effect.

The addition of fat to the diet can increase the supply of long chain fatty acids from the digestive tract and increase milk fat content, but maximum levels in the concentrate feed should be 7–8 per cent, because at higher levels microbial activity and VFA production in the rumen are reduced, resulting in reduced milk yield levels. The protection of fats to prevent their breakdown in the rumen allows higher intakes of fat (up to 30 per cent of the concentrate diet has been fed under experimental conditions) with beneficial effects on milk fat content and yield.

Milk protein

A deficit of energy in the diet leads to a reduction in milk protein content, but an excess of energy will only lead to slight increases. Average levels of milk protein are 3.3 per cent. Cows in early lactation in negative energy balance thus have depressed milk protein contents which probably arise from a deficiency in the supply of microbial protein from the rumen.

Feeding additional protein to cows on an unusually low protein diet (less than 13 per cent crude protein in the DM) will lead to an increase in milk protein content due to an increased microbial protein supply, and in undegradable protein which by-passes the rumen. At higher dietary levels of protein, the excess ammonia produced in the rumen will increase blood urea levels and the non-protein nitrogen content of milk.

For cows producing high milk yields, dietary protein with a low degradability is required to maintain milk protein content. This can be achieved either by offering feeds of inherently low degradability such as fish meal, or by reducing the degradability by heat treatment, or by protection with chemicals such as formaldehyde.

Milk lactose and ash

Changes in food intake have only marginal effects on milk lactose content, which averages 4.7 per cent. A low energy intake can, however, reduce the level by 0.1–0.2 percentage points.

The ash content of milk varies very little with different feeding levels and averages 0.7 per cent.

Strategy of feeding

Feeding the dairy cow involves a complex management strategy of producing adequate grass and forage of the desired quality, for the desired number of cows, and at the same time optimizing the level of concentrates.

The philosophy of feeding dairy cows has traditionally ranged from the self-sufficiency approach typified in the 1950s and 1960s by the successful systems of Rex Paterson, to the high concentrate/high milk yield approach of Robert Boutflour. The former approach attaches great importance to the intensive utilization of grass and low concentrate inputs, in contrast to the major objective of high milk yields per cow of the latter.

These apparently polarized philosophies have each stood the test of time, and are each as vehemently supported by their followers, because the three major factors associated with feeding management which control profitability can be combined in different ways to produce similar levels of profitability. The factors are:

1. Annual milk yield per cow.
2. Annual concentrate input per cow.
3. Annual stocking rate.

The interrelationship of these factors for different levels of grass/forage utilization is illustrated in Fig. 4.10. This represents the requirements for concentrates, not the responses to concentrates (1 tonne of concentrates is equivalent in energy to about 2 000 kg milk, whereas the response to 1 tonne of concentrates is only about 1 000 kg milk; the remainder of the energy from the concentrates increases liveweight gain and/or substitutes for grass/forage intake).

By applying the prevailing milk and concentrate prices to these relationships, projected margins for different combinations of milk yield, concentrate input and stocking rate can be calculated. In the example in Table 4.8, although the high concentrate system has a lower margin per cow, it has a greater margin per hectare due to the higher stocking rate. However, the additional fixed costs associated with keeping additional cows will also have to be included on the high concentrate system.

An important aspect in choosing an appropriate concentrate input and stocking rate is the amount of homegrown feed utilized per hectare. Using previous records of milk output per cow, concentrate input per cow

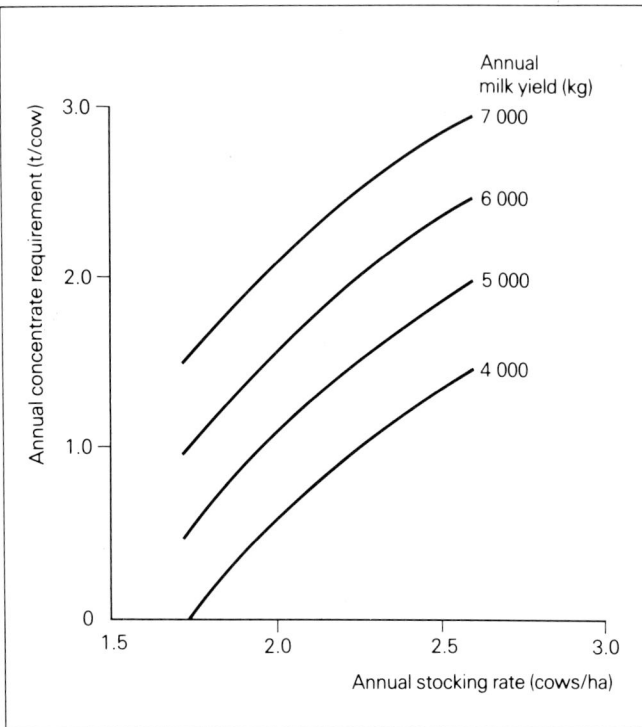

Figure 4.10 The relationship of annual stocking rate, annual milk yield, and annual concentrate requirement (based on a grass utilization of 7t DM/ha)

Table 4.8 Examples of a low-and high-concentrate system

	Low concentrate	**High concentrate**
Annual milk output (kg/cow)	5 500	6 500
Annual concentrate input (t/cow)	1.0	2.5
Annual stocking rate (cows/ha)	1.96	2.60
Grass utilization (t DM/ha)	8	8
Relative M/C*		
per cow	100	85
per ha	100	113

* M/C = margin of milk price over concentrate cost

and annual stocking rate, the level of utilization can be approximately estimated using the utilized ME method (Table 4.9). The average level of grass and forage utilized on farms is only about 6t DM/ha with a range for different farms of 4–11 tonnes. The reasons for low levels of utilization include: low fertilizer input, low rainfall, a short growing season in some areas, poor utilization in the grazing season and large DM losses during ensiling.

Table 4.9   Calculation of utilized metabolizable energy and dry matter per hectare

Example
  Annual milk output 6 000 kg/cow
  Annual concentrate input 2 t/cow
  Annual stocking rate 2.45 cows/ha

	('000 MJ of ME)
Maintenance requirement 365 days × 60	21.9
Milk output 6 000 kg × 5.0	30.0
Growth and pregnancy	3.5
Total requirement per cow	55.4
Minus concentrates 2 t × 10.750	21.5
	33.9
Utilized ME/ha (ME per cow × 2.45)	83.1
Utilized DM/ha (÷ 10.75)	7.73 t

Source: MAFF, DAFS, DANI (1975), *Energy allowances and feeding systems for ruminants*, Technical Bulletin 33, HMSO, London

Winter feeding management

5

The 'winter' period of feeding when grazed grass is not available ranges from 3 months in parts of the extreme south-west to about 8 months in some upland areas of Britain. The provision of adequate amounts of high quality forage is therefore an essential part of management, which can often have a greater impact on the profitability of dairying than the actual feeding system employed. The daily amount of forage dry matter eaten during this period generally amounts to over 1.5 per cent of liveweight.

Forage production

Types of forage

Traditionally, hay and, in some cases turnips, were the staple forage for dairy cows, but since the 1950s, as herd sizes have considerably expanded, grass silage has become the dominant forage on dairy farms. This is due to its being less dependent on the weather, to its ease of mechanization, and to its adaptability for different methods of storage and feeding. However for smaller herds (under 50 cows) hay continues to be widely used either as the whole or as part of the forage diet. In arable areas, silage made from maize, barley and wheat, and root crops, also contributes slightly during the housing period. Examples of the composition of different forages are given in Table 5.1.

The choice of forage is thus mainly dictated by the size of

Table 5.1 Chemical composition of various forages*

Forage	DM (%)	Analysis of DM (%)						
		crude protein	ether extract	crude fibre	NFE	total ash	D value	ME (MJ/kg DM)
Grass silage (VG)	20	17	4	30	39	10	67	10.2
Grass silage (mod.)	20	16	4	34	37	9	58	8.8
Maize silage	21	11	6	23	54	6	65	10.8
Barley silage	40	10	2	25	57	6	62	9.6
Wheat silage	40	8	2	30	56	4	55	8.4
Grass hay (VG)	85	13	2	29	47	8	67	10.1
Grass hay (mod.)	85	9	2	33	50	7	57	8.4
Swedes	12	11	2	10	72	6	82	12.8
Turnips	9	12	2	11	67	8	72	11.2
Kale	14	16	4	18	49	14	69	11.0

* *Source*: MAFF, DAFS, DANI (1975), *Energy allowances and feeding systems for ruminants*, Technical Bulletin 33, HMSO, London

the farm, the rainfall and the availability of alternative forage sources such as on arable farms.

Amount of forage

The amount of forage required for the winter feeding period depends on whether it is offered *ad libitum*, the duration of the winter, the amount of concentrates (or other feeds) offered, and the quality of the forage.

Feeding restricted amounts of forage is not desirable as it will result in reduced levels of animal performance, particularly amongst the less competitive cows in the herd. For cows offered *ad libitum* forage, an indication of the amount required to be ensiled is given in Fig. 5.1.

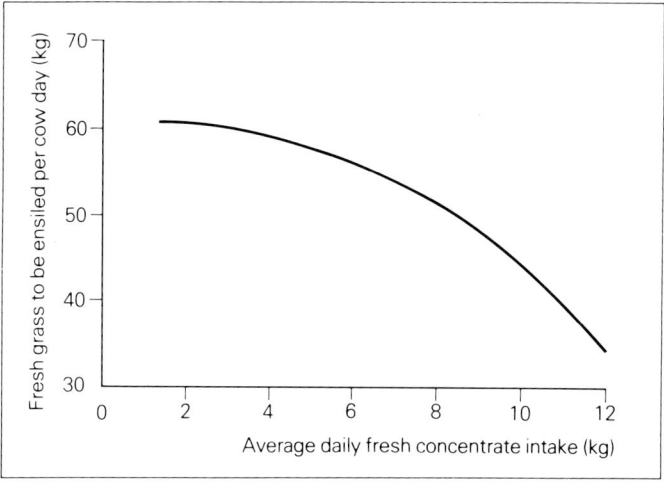

Figure 5.1 The amount of fresh grass (25 per cent dry matter) to be ensiled per cow day for different winter concentrate feeding levels (600 kg liveweight)

The achievement of high yields of silage dry matter per hectare requires attention to a number of basic rules:

1. Good drainage.
2. Adequate lime status of the soil (pH over 5.8).
3. Productive species and varieties of grass in the sward and freedom from weeds.
4. Optimum nitrogen fertilizer application (this will depend on climatic factors and soil nitrogen contribution but for most farms will be in the region of 250–450 kg N/ha).
5. Optimum phosphate and potash fertilizer application (this will depend on the soil status of these nutrients and the amount of slurry/farmyard manure applied but will be about 60–80 kg/ha for most swards).

If these rules are observed then yields of herbage DM will range from 7 in areas with low rainfall to over 13 t/ha in high rainfall areas with a long growing season. If the limitation of water is removed by the introduction of irrigation the lower level will be lifted to over 10t DM/ha.

Silage quality

Taking cuts of silage earlier in the season and cutting more frequently leads to increased digestibilities of the forage. Unfortunately this also leads to a reduction in the amount of DM produced (Fig. 5.2). In general, digestibilities of at least 65 D (digestible organic matter in the dry matter) are needed for high milk yields, but in later lactation and in the dry period lower values are adequate (down to about 58 D). The first cut of silage normally contributes well over 50

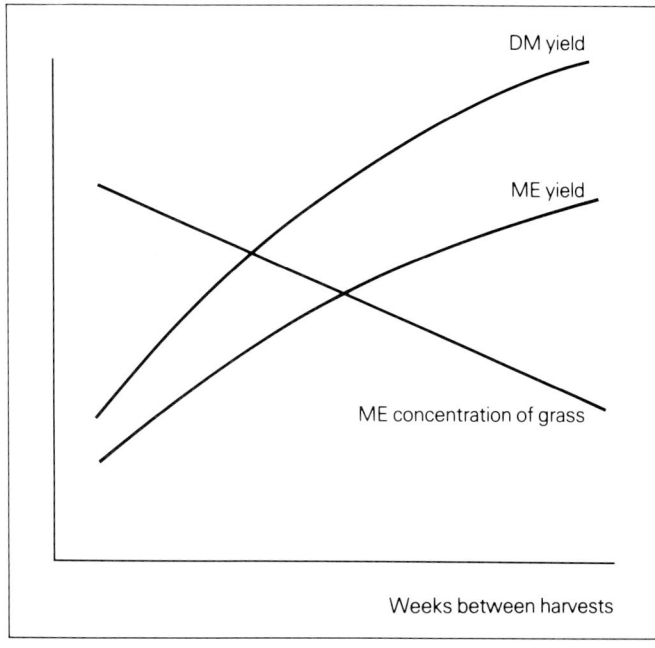

Figure 5.2 The relationship between cutting frequency and yield and quality of grass

Table 5.2 An example of three-and two-cut silage systems to produce 1.7 t DM/cow

Cut	3-cut system				2-cut system			
	Date	ha	Yield (tDM/ha)	Total yield (tDM/ha)	Date	ha	Yield (tDM/ha)	Total yield (tDM/ha)
1	25/5	0.23	4.5	1.04	4/6	0.21	6.0	1.26
2	6/7	0.15	3.0	0.45	23/7	0.15	3.0	0.45
3	17/8	0.07	3.0	0.21				
Total		0.45		1.70		0.36		1.71

per cent of the total dry matter conserved and therefore has a major influence on overall quality. Subsequent cuts have lower DM yields and are taken from a conservation area which reduces in size during the season as the demand for grazing area increases (Table 5.2).

Wilting of silage

The main reasons for wilting herbage prior to ensiling are: to reduce effluent production, to improve the silage fermentation obviating the need for silage additives, and to improve the feeding value.

Silage effluent production is negligible or absent above 23–25 per cent DM content, and where no provision is made for the collection of effluent, wilting is necessary to prevent its subsequent pollution of watercourses. If, however, the effluent can be handled, there are advantages in not wilting, because the system is then less weather dependent. Also, although there are effluent losses of DM due to not wilting, this is compensated for by reduced DM losses in the field compared with wilted grass.

In the ensiling process the intra-cellular plant juices are released and their water-soluble carbohydrates are anaerobically fermented by lactobacilli to produce mainly lactic acid giving the silage its normal pH of 3.7–4.3. If wilting is carried out, the fermentation is restricted and the resulting silage has a higher pH (normally over 4.1), a higher water-soluble carbohydrate fraction and a lower volatile fatty acid content (VFA) content (Table 5.3). The wilting process also inhibits the growth of *Clostridia* which produce low quality unpalatable silages with a high butyric acid content. It is therefore possible to have stable wilted silages at pH levels of over 5.0. However, it is important particularly with heavily wilted silages (over 35 per cent DM) that air is excluded both during the ensiling process by rolling and sealing, and during emptying, otherwise aerobic degradation takes place. This results in heating, moulding, high DM losses, and reduced nutrient value and palatability.

Table 5.3 Chemical analyses of early-cut unwilted, wilted and unwilted + formic acid silages

	Unwilted	Wilted	Unwilted + formic acid
DM (%)	19.4	30.2	19.6
% in DM			
Crude protein	14.0	14.3	13.8
Crude fibre	31.7	30.2	31.5
Ether extract	3.7	4.0	3.6
Ash	9.6	10.3	9.6
Lactic acid	3.6	2.7	4.8
Acetic acid	3.1	0.7	0.9
Soluble carbohydrate	1.1	3.3	1.5
D value	67.2	66.8	69.2
pH	4.9	4.5	3.9
Ammonia N % of total N	16.8	10.4	5.6

Wilting is only partially effective in preventing proteolysis (protein breakdown) during the ensiling process, and even in well-made silages, the protein nitrogen expressed as a percentage of total nitrogen is reduced from about 65 per cent in the crop to about 25 per cent in the silage. However, volatile nitrogen products do decrease as the DM content increases indicating the prevention of breakdown of the amino acids.

The digestibilities of both organic matter and nitrogen of wilted silages are generally 1–3 units lower than for unwilted silages. However, DM intakes in lactating cows fed concentrates are about 10 per cent higher than for unwilted silages with no additive, although this is not always reflected in increased milk production. Where unwilted silages are well preserved by the use of a suitable additive, cow performance is equally good if not better than for wilted silages.

Chopping silage

Different types of forage harvester produce silage of different chop lengths, and this is important in its effect on silage fermentation and cow intakes.

The flail-type and double-chop machines are the simplest harvesters and can be used both for direct cutting, and for picking up from the swath. The chop length tends to be variable averaging over 100 mm for the lacerated material produced by the flail-type and over 75 mm by the double-chop.

The precision chop harvester (also called metre chop or full chop) is a more complex and therefore more expensive machine which picks up the grass from a cut swath, chopping the herbage to a more uniform length of 5–75 mm which varies according to the setting. It requires more tractor power than the simpler types of harvester and is more susceptible to damage by stones and metal objects picked up from the swath. It has a high work rate with moderate sized machines picking up 15–25 tonnes fresh herbage per hour, larger machines at 25–35 and self-propelled models at over this rate.

On smaller farms, forage wagons are an alternative option to the forage harvester plus trailer(s) systems. These are lightweight trailers which incorporate a pick-up/chopping attachment. This system is most suitable where there is a one man operation and where high-DM silage is produced. A more recent development is the ensiling of big-bales of grass in bags.

The chopping of silage prior to ensiling allows greater consolidation of material both in the trailer and in the silo. This leads to more favourable anaerobic conditions for the fermentation process, and the chopping mechanism also releases the sugars for fermentation from the herbage. The intake of dairy cows increases as the chop length is reduced, and increases of 10–15 per cent can be expected for precision-chop (under 50 mm) compared with flail-or double-chop material.

Additives for silage

Additives are added to harvested herbage to assist the silage fermentation process, to give some guarantee that a stable and predictable silage will result. With grasses of a satisfactory water soluble carbohydrate content such as perennial ryegrasses wilted to at least 25 per cent DM, additives are not very beneficial. Their most important use is with silages of low DM and low sugar content. If additives are not used with such silages, clostridial-type fermentation results producing a butyric acid-type fermentation, a high pH and a reduced D value. The use of an acid-type additive as in Table 5.3 leads to a lower pH, through encouraging a lactic acid fermentation, a lower fermentation temperature and a more stable silage.

There are four main types of additives:

Acids. The most common is formic acid which has a recommended rate of application of 2.0–2.5 litres/t of fresh herbage (85 per cent formic acid in the additive). Other acids include the organic acids acetic and propionic and their derivatives and the mineral acids sulphuric and phosphoric. These are normally marketed as mixtures in conjunction with other chemicals.

Preservatives. Formalin and related compounds are used partially to protect the herbage from fermentation, thus resulting in higher water-soluble carbohydrate contents, and protein nitrogen contents. The level of application is critical as too little can lead to an unstable silage and too much can reduce voluntary intake. Most preparations also include an organic or mineral acid to overcome these problems and have application rates of 2.5–5.0 litres/t.

Fermentable substrate. A carbohydrate source such as molasses can be used to provide a readily fermentable substrate for bacteria, as well as providing a supplementary energy source in the silage. The main limitation is the difficulty of uniform application of the necessary quantities (9–15 kg/t).

Bacteria. To encourage the rapid rise in lactic acid producing bacteria, additives have been developed which in powder form include species of *lactobacilli*. It is claimed that these result in a more rapid fermentation and reduction in pH, but there is little evidence to suggest that the numbers of bacteria in such additives are likely to have a significant effect on the silage fermentation.

The use of suitable additives such as acids and acid/formalin mixtures in conjunction with ensiled herbage of low dry matter and/or low sugar content leads to increased DM intakes and animal performance.

Silage storage

The main objectives for a silage store are: to produce a satisfactory silage with minimal dry matter losses, and (where self-feeding is practised) to provide a suitable feeding face.

In addition to the field losses of DM of up to 10 per cent ensiling losses can amount to 5–10 per cent during the fermentation process, 0–10 per cent from effluent and 5–10 per cent from respiration during filling and emptying. In poorly managed silage systems, therefore, total DM losses can be 20–40 per cent of the amount originally cut.

To reduce losses to 10 per cent or less during the ensiling process, wilting to 25 per cent DM, rapid filling, rolling and air-tight sealing are necessary prerequisites. The types of silo – tower, bunker, pit and whether roofed or unroofed – have in contrast much less of an effect on the ensiling process providing the management is good.

Choice of silage-making system

The increasing size of dairy farms and the swing to silage systems has meant an increasing silage hectarage. The choice for the farmer is:
(a) to equip himself with adequate machinery;
(b) to co-operate with neighbours in sharing machinery;
(c) to use a silage contractor.
The most expensive choice is (a), but this does have the advantage of greatest flexibility in choosing when, and when not, to conserve grass. Co-operation with neighbours can cause problems if they require the machinery at the same time, but it is often the lowest-cost alternative. The use of silage contractors gives less choice in the timing of harvesting and many contractors refuse to use silage

additives due to their adverse effects on silage machinery. However, contracting has the advantage of rapid harvesting due to the large machines used.

The trend is likely to be towards more direct cutting which is less dependent on the weather and which, combined with the use of suitable additives, should ensure a good silage fermentation. The effluent produced as a result of this approach will have to be stored and used as a fertilizer or incorporated into dry feeds as a supplement.

Forage feeding

Silage self-feeding

The system of self-feeding is ideally suited to herds of up to 120 cows, the main attraction being its low cost, low labour requirement and therefore its simplicity. The cows graze the face of the silage and it is essential to have a fence or barrier to control the feeding and to have a feed face width proportional to the number of cows. The height of the feed face should not exceed 2 m, otherwise a top layer will have to be physically or mechanically removed to prevent collapse of the face.

Where access to the silage face is unrestricted in time which gives over 20 hours per day, a feed face width of 150–200 mm per cow is adequate. If limitations in time are imposed, however, or if low levels of concentrates are fed, this should be increased, although widths of over 300 mm/cow can lead to spoilage of the silage due to aerobic degradation, particularly with dry silages.

A variety of fences and barriers are used to control feeding (Fig. 5.3), the most common being a single strand of wire about 800 mm from the ground, electrified by a fencer unit. The distance from the feed face is maintained at a distance of 400–600 mm if *ad libitum* access is required. Many farmers with limited silage move the wire daily according to silage availability, e.g. a 30-m silo length for a 200-day winter = 150 mm per day. Other types of feeding control include electrified pipes and tombstone barriers.

Rates of eating silage range from 25 to 75 g DM/minute, faster rates being associated with short-chopped and less consolidated material. The time spent feeding varies mainly with the concentrate input, but on average about 180 minutes/day are spent at the feed face in about six visits.

Simple conveying systems

The simplest silage feeding system, involving the removal and transportation of silage to a feeding point, is the fore-end loader. This system has the disadvantage of allowing air to enter the silage face during emptying, resulting in aerobic degradation of the silage with detrimental effects on silage intake and increased DM losses. A slight improvement is obtained by having a grab on the loader. The best method, however, is the

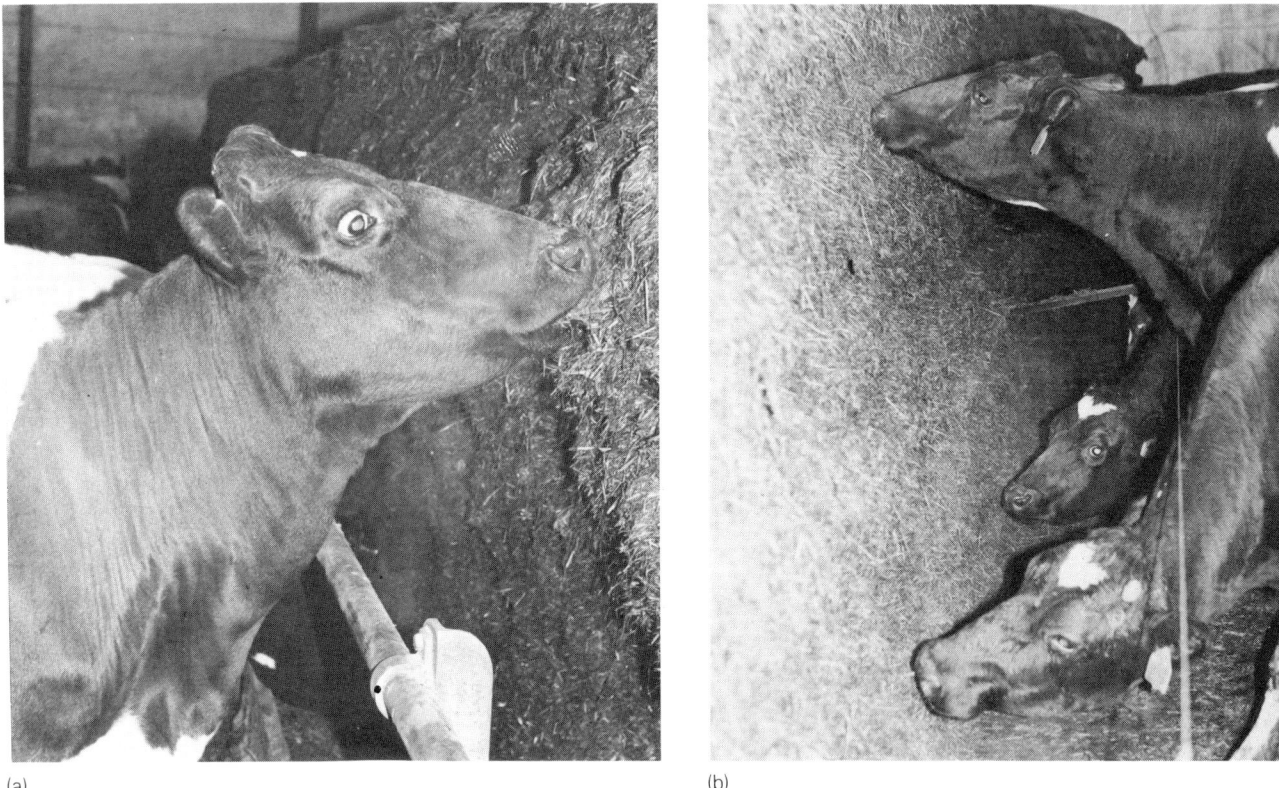

(a)
(b)

Figure 5.3 The self-feeding of silage controlled by a barrier (a) or an electrified wire (b)

Figure 5.4 A block-cutter used for removing silage from silos

Figure 5.5 A forage-box with load-cells to weigh the silage fed to dairy cows in a feeding passage or trough.

tractor-mounted block-cutter which leaves a smooth air-tight silage face (Fig. 5.4). These blocks weighing 400–600 kg can be put in feeding-passages or ring-feeders for easy feeding. They can also be put into forage-boxes for transportation and distribution.

Forage-box feeding

The main advantages of forage-box systems are that they are more adaptable than self-feeding to a range of feeding situations for different groups of cows and youngstock, and

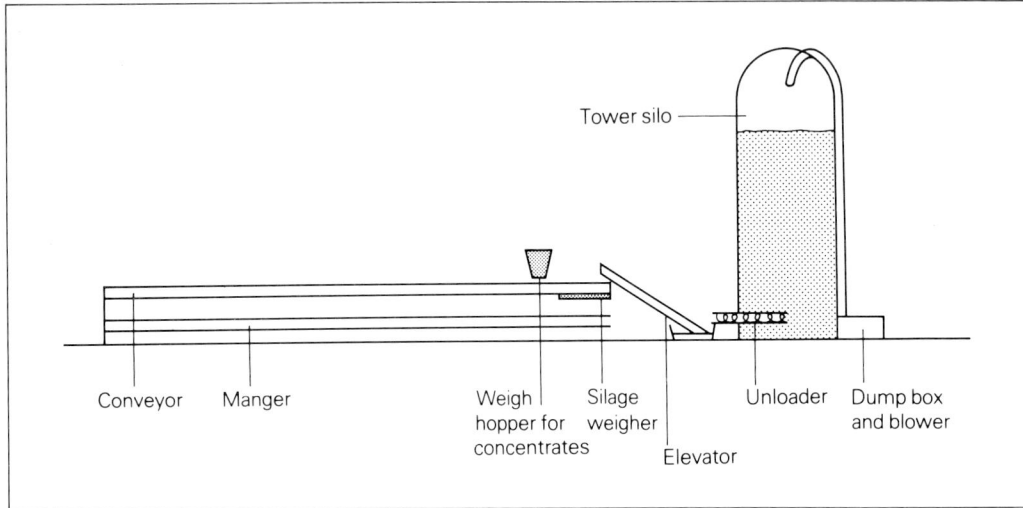

Figure 5.6 A fully mechanized system for silage storage and feeding

Conveyor   Manger   Weigh hopper for concentrates   Silage weigher   Elevator   Unloader   Dump box and blower   Tower silo

in models with load cells, measured amounts of forage can be fed (Fig. 5.5). Also other forages and concentrates can be included with the silage. The DM intakes of silage fed from a forage box are similar to those achieved with well managed self-feeding, although in practice greater intakes than on self-feeding are often achieved due to inefficient control of the self-feed face.

The disadvantages of the system compared with self-feeding are the additional capital cost of the forage box, loading equipment and feeding passage, the greater labour requirement, and the cost of maintenance and repair. The system is mainly justified, therefore, where self-feeding is not possible, and where there are large numbers of cattle.

Mixer-wagon feeding

The complete diet system involves mixing forages and concentrates in a mobile mixer-wagon which dispenses measured amounts of the mixture in a feeding passage. The mixing of the complete diet prevents the cows from selecting individual ingredients from the feed.

Mixes with a higher energy concentration (11.5–12.0 MJ/kg DM) are normally fed to high-yielding cows (ratios of concentrate DM : forage DM of up to 75 : 25), with lower concentrations (under 11.0) to later-lactation cows and dry cows. In some herds a constant mix of about 50 : 50 concentrate DM: forage DM (11.0–11.5 MJ/kg DM) is fed to all cows.

A further advantage of the system is that the mix stores well with little aerobic degradation for 2–3 days and consequently if large-enough mangers are available, the feeding frequency can be reduced to two or three times per week.

Mechanized feeding

Fully-mechanized systems with low labour requirements normally involve tower silos for the storage of silage, conveying systems to mangers and the facility to weigh the silage and add weighed amounts of concentrate feeds during the conveying process (Fig. 5.6). Such systems have a high capital and maintenance cost which has prevented their widespread acceptance. Also the silage for towers has to be over 30 per cent DM for safety reasons and there are difficulties in achieving this without undue field losses in many dairying areas.

Trends in silage feeding systems

The self-feeding of silage has proved a satisfactory method of feeding since the 1950s and is likely to continue as a low labour-input system for groups of cows of around 80 to 100. In practice its disadvantages are: the poor performance which is sometimes seen with young heifers; choosing an adequate controlling system for the feed face which balances low silage wastage with high silage intakes; the need for additional out-of-parlour concentrate feeding facilities if high-concentrate levels are fed.

A feeding passage serviced either with silage blocks, a forage box or in some instances a mixer wagon is likely to continue as the most popular silage feeding system. It has the advantage (providing 650 mm/head is available) of causing less competition between animals for feed space, and allowing the out-of-parlour feeding of concentrates, either by hand or by incorporating the concentrate feed with the silage in the forage box. It is therefore a more flexible feeding system than self-feeding, but has a greater capital cost, and has greater labour and mechanization costs.

The fully-mechanized systems of tower silage and conveying systems are unlikely to increase in importance due to their need for dry (over 30 per cent DM) forages which are difficult to produce in the main dairying areas. The high capital cost and the need for specialist knowledge in mechanization are additional obstacles to their adoption.

Other forages

Hay is declining in popularity as a dairy-cow feed, partly because of its poorer nutritional content compared with silage, but also because the conventional conservation system producing the hay bale is not suited to large dairy units. The conventional small bales were designed for more labour-intensive situations and the big bale also has the disadvantage of being unchopped and the long strands tend to be pulled from the manger into the slurry.

Kale can be a valuable crop for cutting or grazing in autumn or winter. Marrow-stem kale varieties are not winter-hardy and are therefore useful up to the New Year. Thereafter, 1 000 head varieties can be used. The yield can be 6–8 DM/ha for autumn and 4–6 tonnes for winter varieties. Because of poaching and the problem of mud on the cows' udders, direct drilling into paraquat-treated grass swards has advantages. The percentage of the crop utilized, however, is unlikely to be more than 80 per cent. To increase the utilization, the crop must be cut with a forage harvester and carted to the housed cattle. Kale can cause an anaemic condition in cattle if it makes up too high a proportion of the total diet, and the normal recommendation is that it should constitute no more than one-third of the total DM intake.

Root crops fed in conjunction with hay were traditionally fed to dairy cows. Whilst this regime has tended to disappear, stubble turnips (Dutch White) sown into a cereal crop before harvest, or into the stubble after an early harvest, can provide a useful grazing crop for dry cows. The utilization is likely to be only 50–60 per cent of the total DM yield of 4–6 t/ha. Swedes and turnips have a low DM content but can provide a useful energy supplement. Although their crude protein content is 11–12 per cent of the DM, almost half of this is in the form of non-protein nitrogen and if fed in conjunction with grass silage, most of this will be excreted. Fodder beets have a higher DM content (14–22 per cent) than swedes and turnips, and their sugar content provides an alternative energy source to the diet. Most of these root crops can be harvested and offered in the chopped form or whole from self-feed hoppers.

Cereal silage made from whole-crop maize, barley, and wheat can, under certain conditions, produce satisfactory DM yields of 6–10 t DM/ha, using much lower nitrogen fertilizer inputs than for grass silage. However, the protein and mineral content is also low. Forage maize is very sensitive to temperature and for this reason it only gives consistently good yields in the south-east of England.

Straw has a limited value as a feed for lactating dairy cows due to its low energy and protein content. The treatment of straw by sodium hydroxide or by ammonia can, however, increase its energy value considerably. The digestibility is often increased by these processes from 40–45 per cent to over 55 per cent. This results in an increase in intake. With sodium hydroxide, the chemical reaction takes place within hours and the treated straw can be fed immediately or ensiled, the pH remaining fairly stable at over 9.0. The treatment of baled straw with ammonia takes a few weeks for the chemical reaction to take place. This technique has advantages over sodium hydroxide treatment because the nitrogen content of the straw is increased, and the high water intakes seen with the latter treatment do not occur.

Concentrate feeding

Types of concentrates

The term 'concentrates' includes a range of raw materials which may be fed singly or in combination. The main types are as follows:

1. 'Straights' – single feedingstuffs such as barley, maize, sugar beet, soyabean meal, fish meal.
2. 'Compounds' – mixture of straights plus minerals and vitamins.
3. 'Protein concentrates' – mixtures of high-protein straights fortified with minerals and vitamins, together with possibly some cereals and cereal by-products. They are designed for mixing with cereals at inclusion rates of up to 50 per cent.
4. 'Supplements' – included at low rates (under 5 per cent) in the total ration and may contain vitamins, trace minerals, pharmaceutical additives in conjunction with a carrier of ground cereal.

The main source of concentrates for dairy cows is compound feeds and these are formulated by the merchant to a given specification of oil, protein, ash and fibre on a fresh-weight basis and these levels have to be declared to the customer by law. In the formulation, other nutritional and processing considerations have to be taken into account and for this reason there are minimum and maximum possible inclusion rates for each ingredient. A computer using linear programming techniques is normally used to formulate least-cost compounds within the stated constraints. Compound feeds for dairy cows normally contain about 70 per cent cereals, cereal by-products and cereal substitutes, 15 per cent high-protein straights, and 15 per cent other ingredients.

The chemical composition of some concentrates is given in Table 5.4. Due to the inclusion of minerals in

Table 5.4 Chemical composition of various concentrates*

Concentrate	DM (%)	Analysis of DM (%)						
		crude protein	ether extract	crude fibre	NFE	total ash	D value	ME (MJ/kg DM)
High energy compound	86	21	7	5	60	7	76	13.2
Low energy compound	86	18	4	7	61	10	74	11.8
Dried grass nuts	90	19	4	21	46	10	68	10.6
Sugar beet pulp	90	10	7	20	66	3	84	12.7
Fresh brewers grains	22	21	6	19	50	5	59	10.0
Barley	86	11	2	5	80	3	86	13.7

* *Source*: MAFF, DAFS, DANI (1975), *Energy allowances and feeding systems for ruminants*, Technical Bulletin 33, HMSO, London

compounds, which reduces their energy density, this has to be compensated by the use of high energy cereals such as maize and the addition of oils either as an ingredient or sprayed onto the cubes.

Concentrate feeding

Concentrates are the most expensive feed offered to dairy cows and therefore control over their feeding is important. Four main methods are practised on farms.

In-parlour. Most milking parlours are fitted with concentrate dispensers (Fig. 5.7) which allocate concentrates in a manger at a level determined by the herdsman. The feeding mechanisms available include augers, moving plates, vibrating plates and tipping trays, the latter type being the most accurate. Regular calibration of dispensers is important because the accuracy of dispensing varies with the type of concentrate fed and this often changes from load to load. The amount of concentrates which can be fed in the parlour depends on the length of the parlour and the work routine. For most parlours, however, about 4 kg per milking is the maximum amount which can be consistently eaten.

Out-of-parlour. Due to limitations in the amount of concentrates which can be fed in-parlour, a number of different types of out-of-parlour dispensers have been developed. These are sited in the housing area and allow selected cows to have additional concentrates. They range from the simple *ad libitum* type where selected cows have a chain around their necks which gives them access to the feeder, to the more sophisticated programmable type where the cows are individually identified from transponders around their necks. The simple type has the disadvantage that greedy cows eat more concentrates than they require, and timid cows do not get enough and in many cases none at all. For this reason no more than 20 cows per feeding point should be recommended and where moderate to high quality forage is available *ad libitum*, only cows with yields over 23 kg/day should be allowed access. The programmable type of dispenser (Fig. 5.8) normally divides the daily allocation into one quarter for each six hour period and also has the advantage of allowing individual refusals to be checked on the control box.

Group feeding.   The most common method of out-of-parlour feeding is to offer concentrates either in a feeding passage or in a trough. This is the simplest and cheapest method and works well in conjunction with in-parlour feeders to top-up to individual requirements. The most important aspect of this method is that adequate feeding space per cow of at least 650 mm per cow is available (Fig. 5.9).

Mixing with the forage.   Both forage-boxes and mixer wagons can be used to feed either a proportion or all of the total daily allocation of concentrates mixed with the forage. This has the advantage of the concentrates being eaten in a number of small feeds which prevents digestive disorders by maintaining a more stable pH in the rumen.

Figure 5.7   Concentrate dispensers in the milking parlour which discharge concentrates through the wall into individual mangers.

Figure 5.8 Out-of-parlour concentrate dispenser sited in the housing area

Figure 5.9 Feeding of concentrates to dairy cows in a feeding passage

Choice of concentrate feeding system

For the majority of farms, in-parlour feeding will continue to form the basis of forage supplementation. It provides not only a flexible means of varying the concentrate levels between individuals, but also a simple method of supplementing cows during the grazing season. There is also an important non-nutritional advantage in that it speeds up entry of cows into the milking parlour which is a high priority in large herds with one man milking.

The simple *ad libitum* out of parlour concentrate dispenser has too many disadvantages to be of major importance, but the programmable type, although more expensive, has a number of attractions. It allows large predetermined amounts of concentrates to be fed little and often to individual cows, and it has a low labour requirement.

The mixer wagon offers few advantages over the forage box for feeding concentrates. The mixing of ingredients does not appear to increase food utilization and the main advantage is the compression of the mixed feed which allows feeding to be less frequent. The forage box which can be layered with forage and concentrates does not produce as good a mix as the mixer wagon, but still allows a 'little and often' feeding regime.

The hand feeding of a flat rate of concentrates in a feeding passage or trough still presents the cheapest option for out-of-parlour feeding for many farmers. In conjunction with-in-parlour feeding it also allows large amounts of concentrates to be fed in controlled amounts in a number of feeds each day.

Feeding programmes

The strategy of winter feeding practised by a dairy farmer depends on a variety of economic and social factors.

As outlined in Chapter 4, the main factors affecting margins per hectare are the milk yield per cow, the concentrate input and the stocking rate. At a particular level of grass utilization (utilized ME per hectare) high concentrate systems give greater gross margins per hectare (though not necessarily per cow) than low concentrate systems at present milk and concentrate prices. How much of this additional gross margin results in additional profit depends on the amount of extra fixed costs (labour, interest charges on additional cows and housing) which are required. The decision will therefore depend on the individual farm circumstances.

The most limiting resource of the farm – land, labour or capital will be a dominant factor. On small farms where land is limited a high concentrate input/high stocking rate/high milk output strategy is the obvious solution, whereas on a large farm, where good quality labour and capital may be the limiting factors, a simple low concentrate input/low stocking rate approach could be the best solution.

An overriding element in this is the inclination of the farmer. If he wishes to produce and sell pedigree cattle, for example, then a high-concentrate approach to produce good milk yield records is essential, whereas if the challenge and interest is to produce milk from grass and forage, then a low-concentrate system may be preferred.

The dairy farmer with a particular overall feeding strategy, having produced a given amount of forage for the winter months, of a particular quality for a given number of cows, has to decide on what feeding programme to practise. This is mainly concerned with whether to purchase more forage, how much forage to feed daily, how much concentrates to purchase, and how to distribute this between individuals.

Forage availability

A most important aspect of any winter feeding programme is that the forage should be offered *ad libitum*. If inadequate amounts of forage have been conserved (see Fig. 5.1, p. 67), then forages or forage substitutes should be purchased to make up the deficit on a daily basis.

Concentrates are approximately twice the cost of forages per unit of ME and, except in seasons of severe forage shortage when prices are high, forages or forage substitutes such as brewers' grains are a better buy than concentrates. It is beneficial to have forages analysed because decisions on concentrate allocation depend very much on forage quality.

Amount and type of concentrate

Although it is impossible to generalize about optimum concentrate levels for all herds, the decision for each herd is critical as the average concentrate input per cow is a major factor affecting profitability. The optimum level for each herd for a particular winter depends on the amount and quality of the forage available and on the milk potential of the cows.

Figure 5.10 illustrates how a greater quantity of concentrates is required for lower quality forages to achieve high voluntary intake levels, although the response in total ME intake per increment of concentrates is greater. In this example over the range 5–10 kg/day of concentrates the response in total ME intake is 5.6 MJ/kg concentrates for the moderate forage (60 D) and 4.6 for the high-quality forage (over 65 D), but over the range 10–15 kg concentrates the respective responses are reduced to 3.2 and 0.6 MJ/kg concentrates. Little response will thus be expected over about 13 kg/day of concentrates fed with a moderate quality silage and 9 kg/day with a high quality silage. This only gives an indication of the maximum amounts to feed and over a 200–day winter a September-calving cow on this basis would be fed on

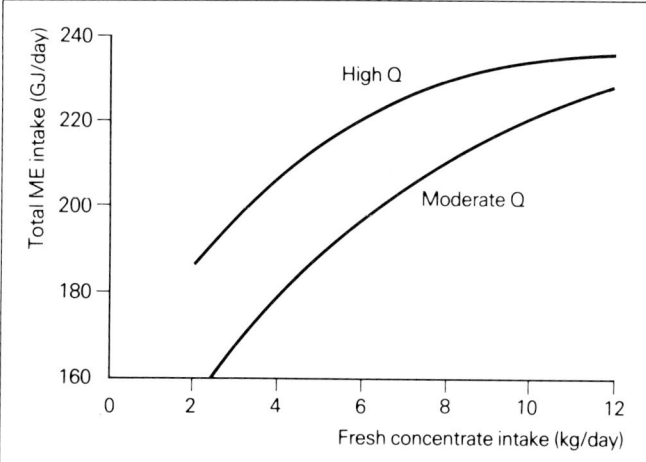

Figure 5.10 Effect of concentrate intake on total metabolizable energy intake when high and moderate quality silages are offered *ad libitum*

average 11.5 and 7.5 kg/day of concentrates for moderate and high-quality silages respectively.

The milk yield potential of a herd is determined by its genetic merit and also by the management imposed. Herds of high potential will show milk yield responses to higher concentrate levels than herds of lower potential. For example, the optimum concentrate level could be greater than the levels indicated in Fig. 5.10 for a herd milked three times daily producing over 7 000 kg milk per lactation, and lower for a herd yielding about 5 000 kg. An indication of potential for individual cows can be obtained from the milk yield at 14 days post-calving, which will be 80–90 per cent of peak yield.

The types of purchased compound concentrates to select from generally fall into the high-/moderate-/low-energy ranges, with crude protein levels (fresh basis) ranging from 22 down to 14 per cent. High energy compounds containing over 13 MJ of ME/kg DM are desirable where high levels of concentrates are to be fed. In general the higher the amount of concentrates fed, the lower can be their crude protein content. The total diet (forage + concentrates) should have a crude protein content in the DM of at least 16, 15 and 14 per cent for early, mid and late lactation, and non-protein nitrogen should not be included in the concentrate portion.

Distribution of concentrates

By comparison with the amount of concentrates fed, the method by which it is distributed between cows and within lactations is of secondary importance. The traditional system of feeding concentrates was to feed according to individual milk yield at a rate of 0.4 kg concentrates/kg of milk. However, as a result of experimental studies in the 1960s and 1970s examining responses to concentrate feeding, alternative approaches are now practised.

In trials where cows were fed hay for maintenance and where different increments of concentrates were fed, greater milk yield responses to additional concentrates were obtained in early than in mid or late lactation, and in cows of high genetic merit compared with those of low merit. However, where cows were fed high quality forage *ad libitum*, the milk yield response to additional concentrates was much less. The reason for this reduced response is that as the concentrate level is increased, the forage intake declines and so the net increase in total energy intake is much less than where the forage is fed at a restricted level. Consequently, although responses of up to 2 kg milk/kg of concentrates have been found with hay fed at maintenance level, with silage fed *ad libitum* the response is often less than 1 kg milk/kg of concentrates, with smaller responses occurring the higher the quality of the forage.

Simple systems of concentrate allocation with *ad libitum* forage have thus been tried with reasonable success and are particularly applicable where low concentrate inputs are practised. More complex systems of allocation relating concentrate input to milk yield are more likely to be used where high concentrate inputs are practised. A whole range of concentrate feeding systems are thus found on farms ranging from the very simple to the very complex. Some of the more commonly used systems are as follows.

Flat-rate system. The development of this simple system has been due to questioning whether it is necessary to vary the concentrate allocation for different cows. The system in its most rigid form involves feeding the same quantity of concentrates to all cows in the herd irrespective of ability and stage of lactation. The system appears to work best in herds where low to moderate levels of concentrates are fed, where block calving is practised and where there is a reasonably uniform herd of cows. The amount of concentrates selected for the herd is normally in the range 4–8 kg/day depending on the amount and quality of silage available. It is essential that in any flat-rate system the silage is offered *ad libitum*, because any restriction in silage availability will limit the performance of higher yielding cows.

If a high level of flat-rate feeding is practised (over 8 kg/day) there can be problems in mid and late lactation of cows becoming too fat and in this case the concentrate level should be reduced from about day 150 of lactation.

Step-system. A modification of the flat-rate system is a step-system, shown in Fig. 5.11. This involves a flat-rate system for about 12 weeks (level depending on silage quantity and quality), followed by a reduction in concentrate input of 1 kg/day for each subsequent four week period. This system has the advantage of feeding additional concentrates in early lactation when the cow may be in energy deficit, and less in mid and late lactation when

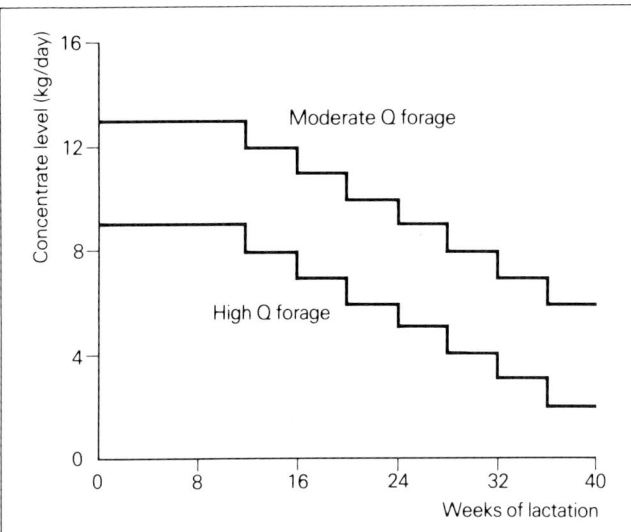

Figure 5.11 A step-system of allocating concentrates to cows offered *ad libitum* forage

Variable-rate. A further refinement of the step-system is to pitch the initial level of concentrates according to the milk yield of the individual cow. One method is to record the yield at 10–16 days post-calving which correlates closely with peak yield (80–90 per cent of peak) and with lactation yield (cows × 250, heifers × 280), and feed *x* kg concentrates/kg milk and then follow the step-system described. The factor will, depend on the desired overall level of concentrates to be fed but will normally be in the range 0.30–0.50 kg/kg milk at 10–16 days.

Complete diets. The mixing of forage and concentrates in a mixer wagon to produce a 'complete diet' ensures that a uniform feed mixture is eaten by the cow with each mouthful. In most dairy herds diets of different energy density are fed to different groups of cows with different yield levels, although in a few a single diet is fed to all cattle. This latter system increases the danger of cows becoming over-fat unproductive and infertile due to over-feeding in late lactation. At high levels of ME concentration (over 11.8) there is also a danger of the cow partitioning a greater proportion of the energy intake to body fat at the expense of milk, as this type of diet leads to the production of a greater proportion of propionic acid in the rumen volatile fatty acids. Feeding complete diets often leads to greater DM intakes compared with feeding the forage and concentrates separately, but there is little

there is an increasing partition of energy to body fat. The system is more appropriate where higher levels of concentrates are fed because it avoids overfeeding in mid and late lactation.

evidence to suggest any improvement in food-conversion efficiency.

The ME concentrations of complete diets are normally in the range 11.4–11.8 MJ/kg DM in early lactation (0–100 days), 11.0–11.4 in mid (100–200 days), and 10.6–11.0 in late lactation and the dry period. It is necessary, therefore, to have at least three yards for the cows unless block-calving is practised. Moving cows from one yard to the next may adversely affect yields for 2–3 days, but has a minimal effect in the long term.

Feeding control

When a feeding programme is put into practice, the performance of the cows may be poorer than expected, due, for example, to an overestimation of forage quality and intake. A monitoring system is therefore an essential part of management.

Monitoring performance levels

The first important measure of performance is the daily milk sales which should be compared with an expected level. If the calving pattern is stable from year to year, then the milk sales graph from the previous year can be used as the control level. If, however, the calving pattern is different, or if the management strategy has changed, a prediction will have to be obtained. A number of organizations provide such predictions which are derived from the calving dates of the cows, their parities and their predicted lactation curves.

In using such predictions, if the actual milk sales are less than the predicted, then possible causes can be investigated, and if nutritional in nature, feeding levels can be increased and the response evaluated (Fig. 5.12).

A more sophisticated system is to collate the data for each month of calving group, by weekly milk yield recording (the monthly recording service provided by the Milk Marketing Boards is too infrequent for feeding-management purposes). The actual average weekly milk yield for each month of calving group is compared (usually a graph) with a prediction or with the yield curve for the previous year, and this will indicate more clearly whether one or more groups are responsible for any decline in milk sales. Suitable remedial action can then be taken.

The body condition scoring of cows is also a useful indicator of the adequacy of feeding. The first critical time to score cows is at about 250 days post-calving because by that time cows should be within 0.5 of the condition score that is required at the next calving. If the score is more than 0.5 under the calving target, additional feeding is beneficial. The second critical time is at about 6 weeks post-calving. If the feeding level is inadequate resulting in the cow being less

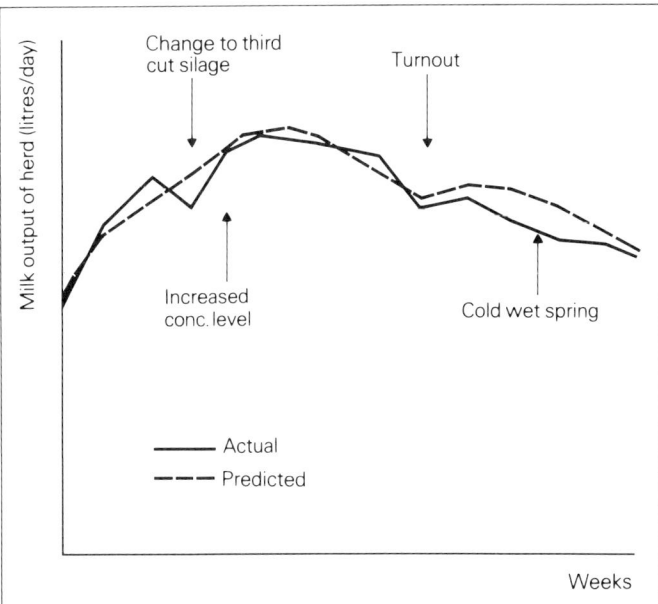

Figure 5.12 The use of a prediction curve of herd milk output, to assist in feeding management

than score 2 at that time, this is likely to lead to reduced milk yield levels, and poorer fertility.

Changes in cow liveweight are difficult to measure accurately due to random changes in gut-fill (eating, drinking, defecating and urinating). This can partially be overcome by automatically weighing cows once or twice daily, and by the use of a computer to estimate liveweight changes with time. However, this information is difficult to interpret for feeding purposes because it very often does not reflect change in fat stores. Heifers and second lactation cows often increase in liveweight from calving onwards due to increasing gut-fill as forage intakes increase and to growth in bone and muscle, whilst at the same time they may be losing body condition which is nutritionally more important. Liveweight *per se* may be important for rationing purposes in a herd, but this can satisfactorily be estimated with a weighband.

A further measure of feeding adequacy in early and mid lactation is the milk protein content. A level of less than 3.0 per cent milk protein in an individual animal indicates underfeeding, i.e. a satisfactory economic response in milk yield can be expected from additional feeding. Levels of less than 2.9 per cent milk protein suggest a fairly severe undernutrition.

Action to remedy poor performance

In cases where the milk yield levels are below the expected or predicted levels, remedial action should be considered.

The average response to feeding an additional 1 kg of

concentrates to cows on·*ad libitum* forage is about 1 kg additional milk. In some circumstances a better response than this will be obtained and in others a worse, but in assessing the response to increased concentrate feeding the effect on cow condition (liveweight) and forage intake should also be taken into account (Table 5.5).

Table 5.5   The response to feeding an additional 1 kg/day of concentrates to cows fed *ad libitum* forage

	Level of milk yield response		
	high	**medium**	**low**
Milk yield increase (kg/day)	1.8	1.0	0.2
Liveweight change (kg/day)	–0.07	+0.03	+0.12
Change in forage DM intake (kg/day)	–0.4	–0.5	–0.6

Cost/benefit = Value of extra milk, liveweight and reduced forage minus cost of 1 kg concentrate

A higher response in milk yield to additional concentrates will occur when:
1. Forage is restricted in quantity (under 90 per cent of appetite)
2. Forage is of poor quality (under 60 D).
3. Concentrates are fed at a low level (under 0.3 kg/kg milk).
4. Cows have a high genetic potential.
5. Cows are in early lactation.
6. Milk protein contents are low (under 3.0 per cent).

If the daily milk sales are graphed and compared with a prediction, or if weekly individual milk yield recordings are carried out, the responses of the herd or month of calving groups within the herd to changes in feeding, can be monitored and assessed. Such systems of feeding control are essential if cows are to be fed at the most economic levels.

Summer feeding management

6

The summer grazing season is the time when milk can be produced most economically, as the cost of grazed grass per MJ of metabolizable energy (ME) is about half the cost of conserved forage and a quarter of the cost of purchased concentrates. Unfortunately on many dairy farms this potential is often not realized and surveys on commercial farms confirm this, showing that on average only 5–6 t DM/ha are utilized compared with a potential for 8–10 tonnes in low-rainfall areas and 10–12 tonnes in wetter areas.

Grazing is a complex management exercise requiring skills both in producing large amounts of herbage and in utilizing it. The interrelationships of the grazing animal and plant are an additional dimension faced by management over those encountered during the winter feeding period.

Herbage production

The amount of herbage utilized per hectare is closely related to profitability and one of the main factors affecting utilization is the amount of herbage produced. This is influenced by a variety of factors, some of which are controllable by management and others which are not.

Climate and soil type

The climate has a major influence on herbage production

particularly rainfall and temperature. There is, however, a strong interaction with soil type. A light sandy soil, for example, will require more rainfall but a lower temperature for optimum growth conditions compared with a heavy clay soil.

Growth of grass in the spring commences when the soil temperature exceeds 4–5 °C, but legumes require a higher soil temperature for growth, and clovers therefore make their main contribution from late May to mid September when soil temperatures are above 9–10 °C.

In the mid part of the growing season the loss of moisture from the sward through transpiration and evaporation, normally exceeds the moisture replenished by rainfall, and soil moisture deficits occur. In low rainfall areas in the east of the country this is the case for most of the season, and as a consequence herbage production is reduced. In areas with large soil moisture deficits, irrigation has to be contemplated. The decision on whether to invest in an irrigation system will depend on the average soil-moisture deficit, the cost and availability of irrigation water and the possible uses for other crops.

The pH of the soil is also relevant, although not as crucial as for most arable crops. It is normally beneficial to apply lime to increase the soil pH to 6.0. At higher pH there is a danger of 'locking-up' trace minerals in the soil such as copper and cobalt leading to reduced levels in the herbage and subsequent symptoms of trace mineral deficiency in the stock.

Herbage species and varieties

A large proportion of the grassland used for dairying can be classed as 'permanent' as it has not been reseeded for at least 10 years. This permanent grass if well managed can be just as productive as temporary leys. The latter are normally very productive in their first full year but thereafter productivity declines, and taking into account the loss of production during the reseeding process the output of temporary grass is similar to that of well-managed permanent grass averaged over 3–5 years. Only in the case of 1–2 year Italian ryegrass (*Lolium multiflorum*) leys is productivity consistently higher than for permanent grass, but this has to be set against the high costs of reseeding.

The major grass species in reseeded leys is perennial ryegrass (*Lolium perenne*), and a wide range of varieties are available with different heading dates and winter hardiness. In theory it is possible to choose varieties with a range of about 4 weeks in heading date so that for conservation purposes a succession of leys can be sown to give high digestibility spread over a number of weeks. In practice most farmers prefer to sow mixtures which are suitable for both conservation and grazing, which include not only ryegrasses with a range of heading dates, but also other species such as timothy (*Phleum pratense*), meadow fescue (*Festuca pratensis*) and clover.

For grazing leys, late perennial ryegrasses which have a more prostrate growing pattern are to be preferred.

Although Italian ryegrass is useful in producing an early grazing about 2 weeks before perennial ryegrass, it is a difficult grass to manage for grazing purposes due to its tendency to produce seed heads throughout the season.

Fertilizer

The application of fertilizer to grassland, supplying nitrogen, phosphate and potash is essential for high levels of productivity.

Nitrogen (N) is contributed to the herbage from a number of sources (Fig. 6.1) which will vary in quantity from farm to farm. On light arable soils the N contribution from the soil may be negligible whereas on long-established grassland it may be over 150 kg/ha. In swards with a high proportion of clover the latter may contribute up to 150 kg N/ha, and on intensive grassland farms a similar amount may be contributed from dung, urine and slurry.

The relationships between fertilizer N input and herbage production is shown in Fig. 6.2. In high-response situations with high rainfall and low soil and clover N, the response is about 30 kg DM/ha/kg N fertilizer. On average the response on most farms is only about half this level. The N fertilizer should normally be spread in fairly equal dressings over the grazing season and the optimum overall level is normally in the range 300–450 kg N/ha (2.0–2.5 kg N/ha/day).

A high proportion of phosphate and potash eaten by the grazing animal is returned in the dung and urine (less than 10 per cent is retained in the body or secreted in milk), although it is not very well distributed over the sward. These nutrients are eventually recycled and in permanent grass swards only 30–40 kg/ha of phosphate and potash are required annually as fertilizer. If slurry is also applied the levels of potash in particular should be closely monitored to avoid the build-up of this nutrient and consequent problems with hypomagnesaemia.

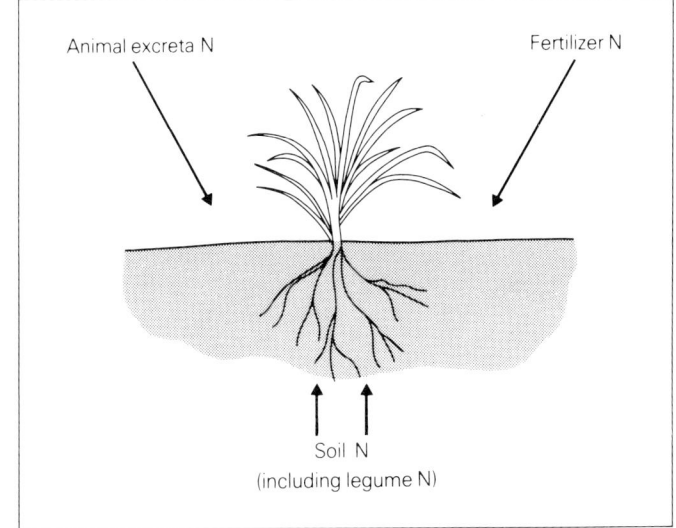

Figure 6.1 Sources of nitrogen for herbage production

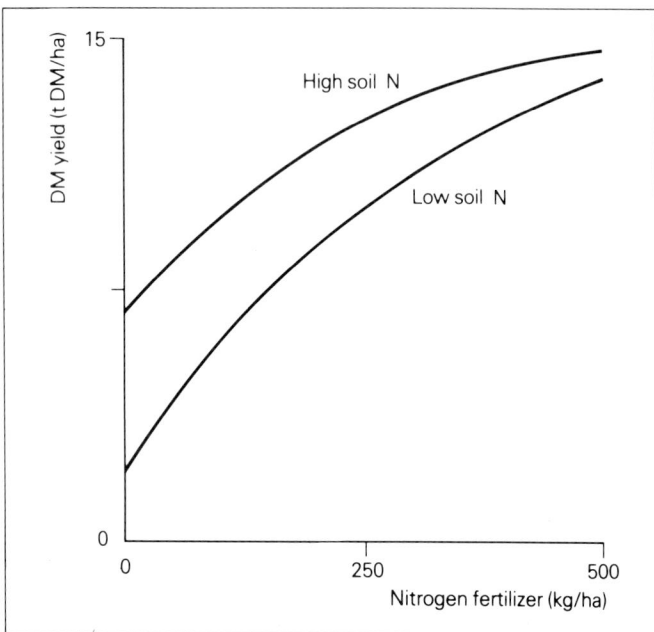

Figure 6.2 The response in dry matter yield to nitrogen fertilizer on low and high nitrogen soils

Effect of grazing

The recycling of nutrients by the grazing animal is one beneficial effect of grazing compared with mechanical harvesting, but grazing generally results in less grass being utilized than with a cutting system.

When conservation cuts are taken for silage, the cutting frequency might be 5–9 weeks, whereas in grazed swards the frequency of defoliation of individual tillers might be 2–4 weeks in rotational systems and 8–12 days in set-stocking systems. It is inevitable that this greater frequency of defoliation leads to a reduced level of herbage production compared with cutting.

An increase in grazing pressure through increased stocking rate also leads to a greater severity of defoliation of individual tillers leading to a delay in regrowth and to a reduced level of herbage production if overgrazing occurs. Part of this effect is due to increased treading, which is more severe under wet conditions.

At the other extreme, undergrazing through understocking of pastures can also lead to a reduction in the amount of herbage harvested. This leads to an increase in the amount of senescence and decay in the sward.

Herbage intake

The milk output per cow from grazed herbage is determined

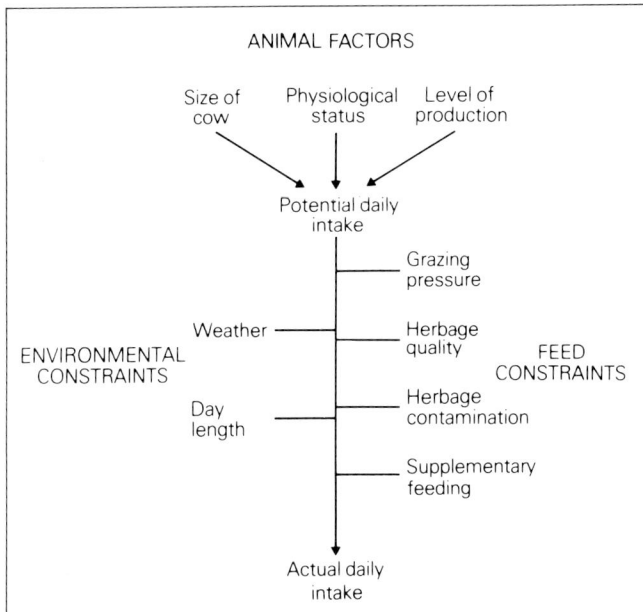

Figure 6.3 Factors influencing the daily intake of herbage

partly by the potential of the cow and partly by constraints associated with the feed and the environment. These constraints determine what proportion of the potential intake is actually consumed (Fig. 6.3).

Grazing pressure

The grazing pressure can simply be defined as:

$$\frac{\text{Number of cows/ha}}{\text{Amount of herbage available/ha}}$$

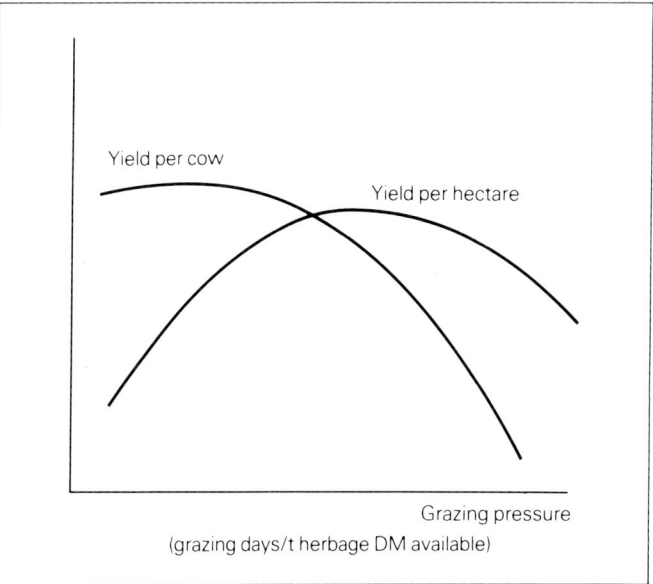

Figure 6.4 The milk yield per cow and per hectare in relation to the grazing pressure

This is probably the most important factor controlling milk output from grass. As the grazing pressure increases, the herbage intake and milk yield per cow decline, but milk output per hectare first increases and then declines (Fig. 6.4). Under rotational grazing conditions (strip or paddock grazing), substantial depressions in intake occur when the stubble height falls below 8–10 cm. In set-stocking systems the critical height of sward in non-rejected areas is about 5–7 cm.

The reason for the depression in intake is that bite size is reduced as grazing pressure increases and the rate of intake of DM is reduced. If the dairy cow is severely underfed she may extend her grazing time, under most systems of management she will graze for only 8–9 hours per day and if she has not eaten her fill in that time, milk yield will be depressed. Thus for high levels of herbage intake, the rate of intake must be high (at least 30g DM/minute).

Herbage quality

There is a linear relationship between the intake of grazed herbage and its digestibility. Unfortunately the initial digestibilities in spring of 75–80 per cent (11.5–12.0 M/D) are not maintained throughout the season. There is a gradual decline in quality as the season progresses even under the best management conditions, and levels of 60–65 D are common in late season. Also if a mid-season drought occurs, depressions in digestibility to 60 D or below may result.

For a continuous supply of high-quality herbage, a high grazing pressure has to be maintained. The only other alternatives are to 'top' the pasture with a mower on one or two occasions during the season, which is wasteful, or to integrate grazing with conservation (alternate grazing and cutting of swards).

Herbage contamination

The herbage available for grazing becomes increasingly contaminated with dung, urine and soil during the season, and rejection of herbage by the grazing cattle increases. It has been estimated that if there was no decomposition, the dung from one cow would cover an area of 80–200 m^2 in a grazing season although this is a maximum value because in addition to climatic effects, decomposition occurs due to microbiological, invertebrate and bird activities. Normally, dung pats decompose in 3–4 months under intensive grazing management. Pasture rejection due to contamination may represent 6–10 times the area of the dung pat but this will depend on the weather (less rejection in dry conditions) and the grazing pressure (less rejection at high grazing pressures). The rejection of herbage around dung pats is thought to be caused by smell.

Supplementary feeding

The supplementary feeding of concentrates to dairy cows during the grazing season is a common practice although trials work has shown that the average response in milk yield is only about 0.3 kg/additional kg of concentrates. The reason for the poor response is the high substitution rate of concentrates for herbage intake, when ample herbage is available.

As the concentrate level is increased, grazing time declines and in many cases the increase in total DM intake is less than 0.2 kg DM/kg of concentrates fed. The extent of this substitution depends mainly on the quality and quantity of the herbage available.

The need for supplementary feeding increases as the season progresses because intakes decline due to the depressing effects of lower quality, increased contamination, and reduced day lengths. The rate of intake per minute declines as the cow becomes more selective and as grazing times are normally limited to about 9 hours every 24 hours, intakes decline. Table 6.1 illustrates how the decline in rate of intake reduces the potential milk yield from grass and unless the cows are moved to clean aftermath, grazing will supply only maintenance plus about 25 kg/day of milk in early season, 15 kg/day in mid season and 5 kg/day in late season. The rate of concentrate feeding should be about 0.45 kg/kg milk above these threshold levels. If shortages of herbage occur, forages such as hay or treated straw should be fed to appetite in preference to feeding additional concentrates, as they have a much lower cost per unit of ME.

Table 6.1 Potential milk yield from grass

Milk yield (kg/day)	DM intake required (kg/day)	Required grazing time (hours/day)*		
		early	mid	late season
5	10.6	6	7	9
15	13.5	8	9	11
25	16.4	9	11	14
35	19.6	11	13	16

* Rates of intake (g DM/min.) decline as the season progresses and cows generally will not graze for longer than 9 hours/day

Environment

The main environmental factors affecting dairy cow performance are the weather and day length. In cold, wet, windy conditions grazing times, and hence herbage intakes, are reduced. To maintain cow performance in such conditions supplementary forages and/or concentrates should be fed. As day lengths shorten, grazing times may also be reduced. On average about 60 per cent of grazing takes place between morning and afternoon milkings, but there is an increase from about 40–75 per cent as the season progresses.

Matching grass utilization to production

In a grazing sward there is a continuous cycle of DM production, utilization by the grazing animal, and senescence and decay of residue (Fig. 6.5). If grazing is inefficient, a longer stubble remains and a greater rate of senescence occurs. As a result less than 75 per cent of the gross DM production of most swards is utilized by the animal.

Measurement of utilization

The amount of grass utilized per hectare can be measured for the whole or part of the season if careful records of animal outputs are maintained. This grassland recording uses the number of grazing days per hectare, the average milk output per cow, the supplementary feed input per cow and an estimate of liveweight change to calculate the utilized metabolizable energy (UME) per hectare. An example is shown in Table 6.2.

Whilst grass-plot trials suggest a possible utilization of over 12 tonnes DM/ha (UME about 135 GJ/ha), in practice on recorded farms only 5–6 tonnes (UME about 60 GJ/ha)

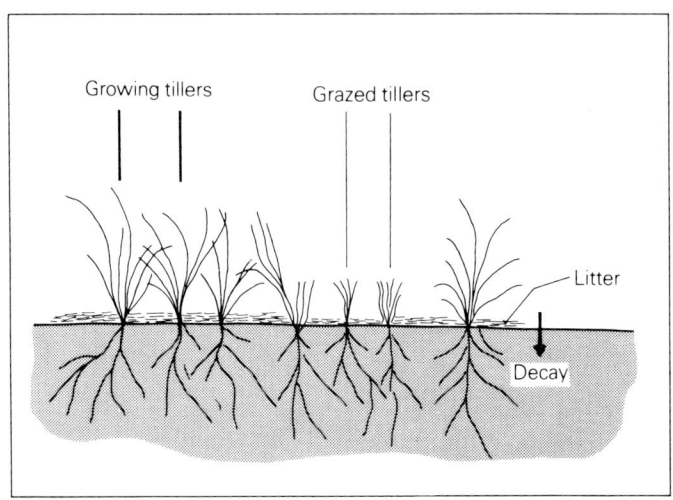

Figure 6.5   The growth, utilization, senescence and decay of grass plants

are utilized. The difference between the plot trial results and on-farm results is partly due to a low DM production (inadequate drainage, fertilizer, lime, etc.) on farms, and partly to underutilization of the grass DM produced. Levels of UME for different periods of the grazing season are illustrated in Fig. 6.6

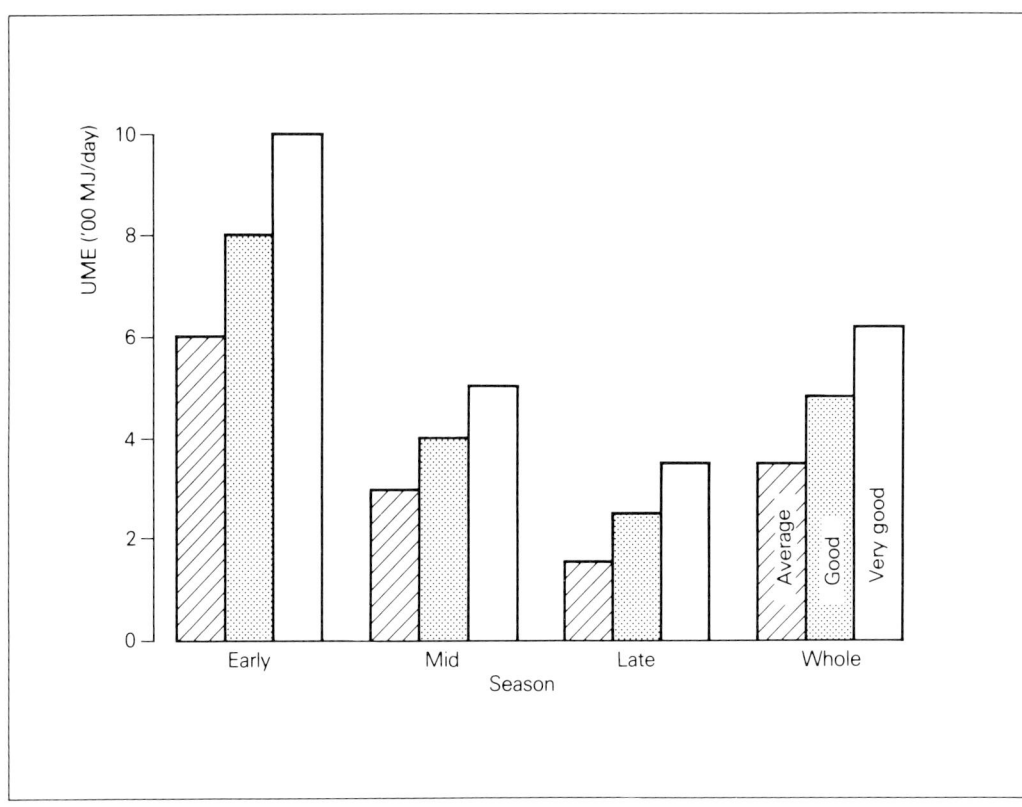

Figure 6.6 Levels of utilized metabolizable energy achieved under average, good and very good grazing management conditions

Table 6.2 Calculation of UME per hectare

Example
160-day season, 528 grazing days/ha,
950 kg milk output/ha, estimated liveweight
increase 304 kg/ha, concentrate input 1 320 kg/ha.

	GJ/ha of ME
Maintainenance – 528 × 60/1 000*	31.7
Milk – 950 × 5.3/1 000	50.4
LW increase – 304 × 34/1 000	10.3
ME output	92.4
Minus concentrates – 1 320 × 10.5/1 000	13.9
UME	78.5

* GJ = MJ × 1 000

Table 6.3 Estimating stocking rate from grass ME output and ME requirements

Season	Liveweight (kg)	Milk yield (kg/day)	LW change (kg/day)	ME* requirement (MJ/day)	Grass ME output (MJ/ha/day)	Stocking rate (cows/ha)
Early	580	23	0.5	201	800	4.0
Mid	610	19	0.5	182	400	2.2
Late	640	15	0.5	164	250	1.5

* Assumes no concentrate feeding. If concentrates fed subtract concentrate ME input from requirement and recalculate stocking rate

Choice of stocking rate

Choosing the correct stocking rate for a particular period of the season is one of the most important decisions that the grazier takes. The choice of stocking rate can be made objectively by matching the estimated grass ME output from Fig. 6.6 to the required ME intake (Table 6.3).

Increasing the stocking rate generally leads to an increase in grass utilization (increased UME), and a reduction in losses due to senescence and decay. However, it may also lead to a reduced milk output per cow which must be closely monitored in case the threshold is reached where milk output per hectare begins to decline (see Fig. 6.4).

Adjustments for grass shortages

Variations in weather conditions between years means that the chosen stocking rate for a particular period may not always be correct for balancing grass utilization with production. This is where 'fine tuning' of grazing management is necessary to deal with short-term grass shortages which can occur in low-rainfall periods.

An indication of whether grass is in short supply can be obtained in three ways:
1. From the daily bulk milk production and regular cow condition scoring. If the rate of decline in milk yield accelerates and particularly if it reaches 2.5 per cent per week, and if cows are not gaining in body condition, then

Figure 6.7 A grassmeter used for measuring the average height of a grazed sward

additional grazing must be found or supplementary feed offered. In herds which are milk-recorded weekly, an even better control can be obtained from the milk yield changes of each month of calving group.

2. From grass availability. The judgement of availability can be made visually from the height of stubble in rotationally grazed areas and of sward height in set-stocked areas. A rough guide is that intakes of grass are reduced at heights of less than 10 cm for rotational and 7 cm for set-stocking areas averaged across the whole field. A more accurate assessment of height can be made with a grass meter (Fig. 6.7). As with (1) above, additional grazing or supplementary feeding should be offered when grass shortages occur.

3. From the intake of a buffer feed. This feed can be silage, hay or straw (the latter preferably treated with ammonia or sodium hydroxide to increase digestibility). If this feed is offered *ad libitum* the cows adjust their intake according to their grass intake. It is therefore a self-adjusting system. The buffer feed should preferably be offered after milking before returning to the field.

Grazing systems

The objectives of an efficient grazing system are to maintain the desired level of milk output per cow, to produce and utilize the herbage efficiently, and to develop a system that is simple to operate.

Rotational systems

The development of intensive rotational grazing systems in the 1950s and 1960s led to a considerable improvement in grassland use. The discipline of applying fertilizer after each grazing combined with the controlled allocation of herbage and the conserving of surpluses has led to an increase in stocking rates. The main systems in use are strip and paddock grazing.

In strip grazing an electric fence is used to allocate areas of grazing which have been growing for 2–5 weeks. The fence is moved once or twice daily and the decision as to how much herbage to offer is based on the amount of stubble left in the grazed area. To prevent the regrazing of stubbles, a back fence is often used if cows are strip grazed in the same field or block for more than 3–5 days (Fig. 6.8).

In paddock grazing, a number of paddocks are formed with electric fencing and these are grazed in rotation. These may supply from 1 to 7 days' grazing depending on their size. A common system is to have about 28 one day paddocks each allowing about 100 cow-grazing days/ha (Fig. 6.9). In early season fewer paddocks are required and some are taken out of the system for conservation, but later in the season all paddocks are grazed. If fewer and larger

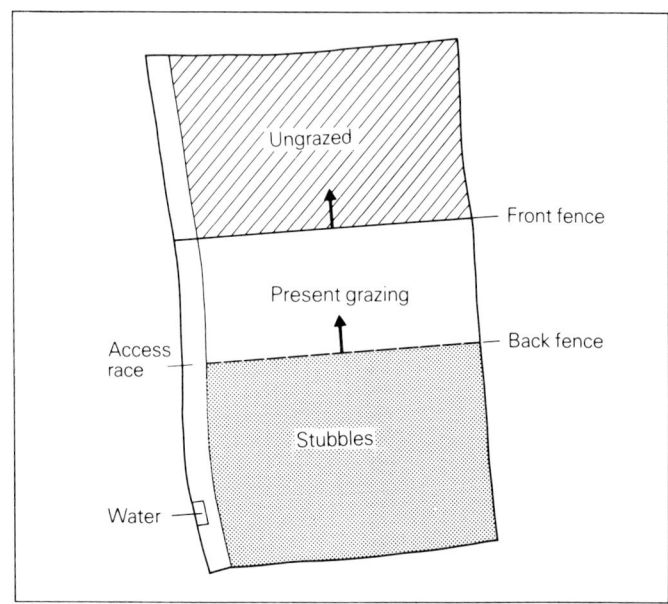

Figure 6.8 A strip-grazing system with once – or twice – daily movement of the electrified front and back fences

paddocks are used where the cows graze each for 3–5 days, similar levels of performance can be achieved to a 1-day paddock system, but greater fluctuations occur in milk output between days. More complex systems with over 40

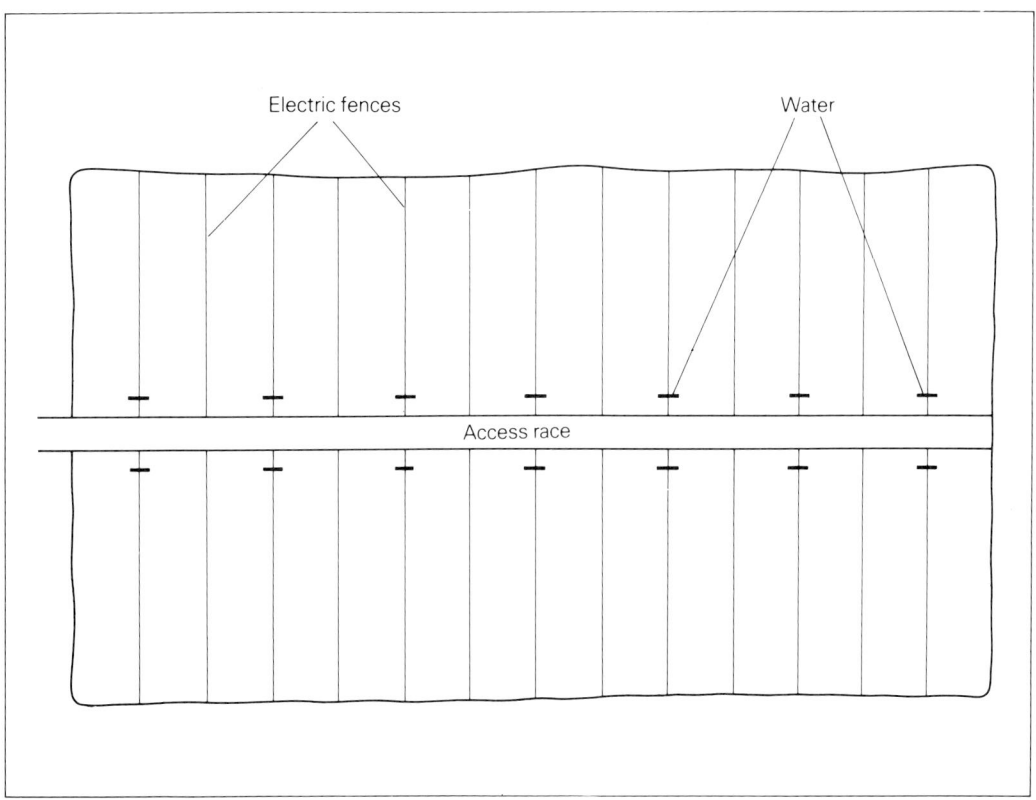

Figure 6.9 A paddock grazing system with 28 1-day paddocks

Electric fences

Water

Access race

paddocks give the necessary flexibility to allow alternate grazing and conservation of paddocks, but the system is expensive and complicated to organize. The benefits in milk output of such systems are only marginal.

A system which allows high yielders to have first choice of the grass is the leader/follower system in which high yielders graze the paddock ahead of low yielders. The decision to change paddocks is made on the basis of the stubble height in the followers' paddock. In practice this system has shown little benefit over the single group systems, and is time-consuming at milking times due to having to move two groups of cattle. The system only has a useful application if dry cows or youngstock are used as followers.

Set-stocking

In set-stocking or continuous grazing systems, the cows are given free access to the whole grazing area. The individual tillers are generally grazed frequently at 8–12 day intervals which could possibly lead to a reduction in herbage production compared with rotational systems. However, this frequent defoliation leads to an increase in tiller population which appears to compensate for the effects of the more frequent defoliations. In rotational systems there is a reduction in tiller population from about 10 000–15 000/m^2 in early season to about half this amount by the end of the season, whereas in set-stocking systems,

the average tiller population is maintained at 12 000–15 000/m^2. As a result little difference in milk output between rotational and set-stocking systems has been found.

The area of grazing has to be increased as the season progresses to account for the reduction in herbage growth rate. In early season stocking rates should be high at 5.5–6.5 cows/ha to encourage tillering from the base of the sward, and to release the maximum amount of land for conservation. If under-stocking is practised in early season, aerial tillering and sward deterioration result. In mid and late season stocking rates should be reduced to 3–4 cows/ha.

The advantages of set-stocking compared with rotational systems are that less fencing and fewer watering points are required. A disadvantage is concerned with the large size of the grazing area, which for herds of over 100 cows can present difficulties in rounding-up the herd for milking. For this reason the set-stocking area is often separated into day and night areas.

The milk yield potential per cow from set-stocking is less than for most rotational systems due to the shorter length of the grass. This reduces the rate of DM intake by the cow and hence the total daily DM intake. The system is therefore less appropriate for high yielding cows (over 30 kg/day of milk) unless high levels of supplementary concentrates are offered.

Zero-grazing

This system involves feeding housed cattle on cut green herbage. It has the advantage of utilizing more DM/ha and thus allows stocking rates to be increased, but has the disadvantages of requiring additional labour and machinery, and of problems associated with housing such as lameness, mastitis, and slurry storage and disposal.

The feeding of cut grass does not allow the degree of selectivity by the cow as for grazed grass and hence DM intakes are generally no greater than for conventional grazing where the latter is unrestricted. Also it is important to cut at least once daily to maintain intakes, as fresh grass soon begins to heat and deteriorate leading to reduced intakes. One method of reducing this effect is to use a silage additive such as formic acid which allows the cutting frequency to be reduced to 2 days.

The greatest benefits of zero-grazing are firstly in the spring where, because fewer hectares are required than for grazing, land can be released for conservation, and secondly in late summer when with conventional grazing, intakes of grass normally decline due to the increase in fouling and reduced day lengths.

Storage feeding

Keeping cows indoors during the summer months and feeding conserved forage is an alternative to grazing. The advantages over conventional grazing are that fewer hectares are required, that greater control over feeding can be maintained and the handling of cattle is easier. Advantages over zero-grazing are the elimination of the daily cutting of grass, and the feeding of a more consistent (though poorer quality) feed. The disadvantages are the need to provide extra silo capacity, and, as for zero-grazing, the additional health problems, and the slurry storage and spreading.

A system which makes the best use of both grazing and storage feeding is to graze the cows at a high stocking rate during the day, and to house them at night offering conserved forage. This acts as a buffer for variations in grass intake during the day.

An example grazing system

On a farm with an average herd size of 100 cows with a grassland area of 40 hectares the grazing system is set-stocking initially on 16 hectares, with an introduction of silage aftermaths as the season progresses (Table 6.4).

Fertilizer is applied in March to the grazing area at a rate of 105 kg N/ha (6 bags/ha of 34.5 per cent N fertilizer). This is followed by subsequent dressings of a grazing compound fertilizer at a rate of 72 kg N, 12 kg P_2O_5 and 12 kg K_2O/ha (e.g. 5 bags/ha of a 29 per cent N, 5 per cent P_2O_5, 5 per cent K_2O fertilizer).

Table 6.4 An example set-stocking grazing system for 100 cows

Period	Grazing area (ha)	Silage area (ha)
1 Mid April – early June	16	24
2 Early June – end July	24	16
3 End July – end August	32	8
4 End August – early October	40	—

A system such as this has a simple framework, and has safeguards built-in to buffer against grass shortages.

The first application is in mid May followed by three similar dressings in June, July and August. For the silage aftermaths used for grazing, the fertilizer applications are 70 kg N, 35 kg P_2O_5, 35 kg K_2O/ha (e.g. 7 bags/ha of a 20 per cent N, 10 per cent P_2O_5, 10 per cent K_2O fertilizer).

The cows are allowed free access to hay each morning after milking for 45 minutes. Average intakes are 1–2 kg depending on hay quality, and vary according to weather conditions and grass availability. If intakes exceed 3 kg/day then access is also given to hay after the afternoon milking. Concentrates are fed according to individual cow milk yields, at a level of 0.45 kg/kg of milk over 25, 20, 16, 12, 9 and 6 kg/day of milk for each successive 4 week period of the season respectively. A concentrate of 14 per cent crude protein (fresh weight) is used with an ME content of at least 12.0 MJ/kg DM. Cows which are dried off during the summer are grazed separately from the milking herd on silage aftermaths.

Dairy herd health

7

The financial losses incurred by dairy cows which are not in a healthy condition are considerable and include:
1. Losses due to death.
2. Losses due to culling.
3. Losses due to reduced production.
4. Losses due to increased veterinary costs.

The annual incidence of deaths is normally less than 3 per cent, and death from old age is a rare occurrence due to the average herd life being only four lactations. The most likely causes of sudden death in grazing lactating cows are hypomagnesaemia in spring and autumn, and hypocalcaemia in cows at parturition. Under housing conditions the most likely causes are injury, severe mastitis (particularly coliform mastitis) or hypocalcaemia at parturition.

The financial losses from culling arise from the difference in value between the sale of a cull cow for meat and the cost of a replacement heifer, plus the value of any deficit in the margin of milk production of the heifer compared with the cow. Also if the culling rate can be reduced through better management, fewer replacements have to be reared, thus releasing land for more profitable enterprises.

Most health problems affect milk production in both the short and the long term, as most problems occur in early lactation. The cost of veterinary treatment can also add substantially to the costs of production.

It is beneficial, particularly with large herds, to maintain records of veterinary treatments. This can be done by the use of a diary or by maintaining individual cow record cards. Such records form a useful reminder for both farmer and veterinary surgeon, about the history of individual cows, and also they can be used for identifying which cows have to be culled.

The two extreme approaches to maintaining herd health are prevention and cure. If the former is carried out satisfactorily, few of the financial losses associated with disease will be incurred, but if the latter is the main approach considerable losses will result. Good preventative medicine is based on having organized and strict routines for cow preventative treatments, examinations and tests. The maintenance of good records is essential together with routine liaison between farmer and veterinary surgeon.

Reproductive problems

Dystokia

In dairy cows the incidence of dystokia (difficult calving) is about 5 per cent and is caused by calves which are large relative to the size of the pelvic opening, or to an abnormal birth position of the calf. In maiden heifers because of their immaturity, the incidence is greater (often over 10 per cent) and depends on the sire of the calf which can significantly affect both calf size and the length of the gestation period.

Cases of dystokia in Friesian and Holstein heifers are quite common when calf birth weights exceed 40 kg.

If a cow is in labour for several hours without the water bag or feet of the calf being seen she should be internally examined. Also if the legs of the calf are seen, and no further progress is made within 2 hours, an internal examination is required. Such examinations should involve strict adherence to aseptic recommendations.

When the calf is in the normal position then assistance should be given by attaching calving ropes to the legs of the calf and by applying appropriate pressure. If the calf is lying in an abnormal position, then veterinary assistance should be sought.

Retained placenta

The placenta is retained after parturition if there is a malfunction of the third stage of labour. A high incidence was associated with brucellosis but since the implementation of the brucellosis eradication scheme, the most common cause is premature birth. Most twin births are premature and about half of these are associated with a retained placenta. There is also a high incidence of retained placenta after early induction of parturition with synthetic corticosteroids or prostaglandins.

The initiation and continuation of contractions in the third stage of labour are under complex endocrine control which is susceptible to upset. Premature births, and normal-time births which are affected by dystokia, hypocalcaemia or bacterial infections (caused by poor calving hygiene) are all predisposing factors to retained placentas.

If the uterus is healthy, injections of oxytocin or prostaglandin in the first 24 hours after birth will release the placenta. If the retention remains untreated at that time it is likely to continue for 6–10 days, and providing it is not associated with infection no adverse effects of waiting for natural expulsion will occur, although there will be some putrefaction. Cases associated with dystokia generally last for 2–3 days and severe endometritis often ensues. If such cases remain untreated, effects on milk production and fertility will occur and severe toxaemia and death are possibilities. An early treatment by the veterinary surgeon of retained placenta following cases of dystokia, is therefore recommended.

Endometritis

Inflammation of the uterus or endometritis is caused by a variety of organisms after parturition. A discharge from the vulva (whites) is seen, the amount and colour depending on the severity of the infection. Prompt attention by the veterinary surgeon is required if milk yield and fertility are not to be impaired.

Poor sanitary conditions at calving, such as dirty calving boxes and the insertion of dirty hands into the vulva of the cow during assisted calvings, are two important predisposing causes.

The prevention of the condition is based on calving in clean conditions (outdoors if possible). If the calvings are indoors, frequent cleaning, disinfecting and bedding are necessary to prevent transmission from cow to cow. The washing and disinfecting of hands, arms and equipment during assisted calvings is also essential.

Delayed onset of oestrus

Routine records of all oestrous periods should be maintained so that non-cycling cows can be detected early and treated by the veterinary surgeon. Most cows exhibit oestrus in the 3–6 week *postpartum* period. However, some cows (in particular high yielders) may show some delay or exhibit irregular oestrous periods. All cows with such abnormalities should be examined and treated where necessary by the veterinary surgeon by 42 days *postpartum*, otherwise prolonged calving intervals will result. It is therefore beneficial to record all oestrous periods after calving so that non-cycling cows can be identified.

Examinations of the reproductive tract may show that many cows are cycling but have failed to be observed in oestrus. Such diagnoses would suggest that increased attention to oestrus detection is required (see Chapter 3).

Low pregnancy rates

In over 5 per cent of recorded inseminations, the cows are not in oestrus and have been wrongly submitted for AI. In cows correctly submitted or where natural service is used over 90 per cent of ova are fertilized. Subsequent pregnancy rates as diagnosed by the veterinary surgeon average 55 per cent. Most of the losses of fertilized ova occur in the first 14 days after service and are characterized by a return to service at about 21 days. Later embryo losses account for about 10 per cent of embryos and are more common in older than in younger cows. Implantation is normally completed soon after 30 days and losses thereafter in healthy cows are small (less than 5 per cent).

If low pregnancy rates are observed, the following possible causes should be investigated:
1. Low bull fertility.
2. Incorrect diagnosis of oestrus.
3. Cows either very thin (under score 1.5) or very fat (over score 3).
4. Inadequate nutrition (energy, protein, fibre, minerals and vitamins).

The diagnosis of a problem is considerably assisted by good records of calving dates, heats, services, veterinary treatments and pregnancy diagnoses. These will show whether the cows are cycling normally, whether AI was at the correct time, and whether most returns to service were to a particular bull.

Metabolic disorders

Hypocalcaemia

This condition – often called 'milk fever' or 'parturient paresis' – occurs within 3 days *postpartum*, occasionally 1 day before, but most commonly on the day of calving. It rarely occurs in heifers and there is a low incidence in second lactation cows. It is associated with the initiation of milk secretion at parturition which requires the secretion into the milk of large amounts of calcium. The incidence is highest in autumn-calving cows at grass and in housed cattle which are 'steamed-up' on compound concentrates.

The disorder is associated with low blood calcium levels which rapidly fall from about 10 mg/100 ml to below 7 mg/100 ml. The first symptom is unsteadiness, and as blood levels fall below 6 mg/100 ml, the cow lies down and eventually becomes paralysed with legs stretched out and often with the head turned towards its side. If treatment is not promptly given then death quickly follows. The cow should be moved to an upright lying position.

Treatment for mild cases involves injecting calcium borogluconate under the skin and in more severe cases injecting intravenously. Many cows relapse some hours later and require further treatment. In cases concerned with calving on spring or autumn grass, the condition can be associated with a sub-clinical magnesium deficiency, and a better treatment response is obtained by injecting a mixture of calcium and magnesium.

A number of preventative measures can be taken which include: (a) feeding low-calcium diets during the late dry period such as barley supplements and introducing high-calcium diets such as balanced concentrates 2–3 days prior to calving; (b) feeding or injecting vitamin D 2–6 days before calving; (c) feeding silage or other forage during the late dry period.

Hypomagnesaemia

The incidence of hypomagnesaemia commonly known as 'grass staggers' is highest in the spring in lactating cows at grass, and can also occur in the autumn. Very often sudden death occurs before any symptoms are seen. In the initial stages the cow staggers around with froth at the mouth, and this is followed by collapse and death. It is caused by a deficiency of magnesium resulting from low concentrations in the grass, and this can be further aggravated by low dry matter intakes. The critical blood level is about 2 mg/100 ml.

The storage of magnesium in the body in a form which can be mobilized is low and therefore the cow depends on an adequate daily intake of magnesium. A long-term preventative measure is to use magnesian limestone whenever liming is necessary.

It is inadvisable to apply potash fertilizer or slurry to grazing land in the spring as this will reduce the magnesium level of the grass. A feed of 60 g/day of calcined magnesite in the concentrate supplement during the first 3–4 weeks after turn-out will normally prevent the condition. In cold/wet conditions in September/October this process should be repeated if hypomagnesaemia is a problem on the farm. The calcined magnesite can also be administered by dusting the pasture at critical times. A further measure is to ensure that cows are eating at least 2 kg DM of long forage (silage, hay or straw).

Ketosis

Ketosis or acetonaemia is a problem of early lactation. The symptoms are a reduced appetite and milk yield, some loss of co-ordination and a sweet smell to the breath and milk.

The ketone bodies (acetone, acetoacetic acid and hydroxybutyric acid) which cause the smell are the products of incomplete mobilization of body fat. The problem is most common in late winter when forages are possibly of poorer quality and in short supply, and in older high producing cows.

The main preventative measure is the maintenance of a high energy intake in the first 6 weeks after calving. Other disease problems such as endometritis and mastitis which cause reduced food intakes can also predispose the cow to ketosis.

The condition arises from insufficient glucose for the metabolism of body fat and one treatment for the condition is to infuse glucose into the blood, although the effect is only transient. In general, ketosis is naturally alleviated when the milk yield and food intake are back in balance, which unfortunately is not until a large amount of potential milk yield has been lost.

Bloat

Bloat is normally caused by the production of a stable foam in the rumen resulting from gas production in the fermentation process. In this foam the gas is trapped and cannot be belched out in the normal way and pressure builds up in the rumen. If untreated, death results from respiratory and circulatory failure.

The condition is most commonly seen on legume diets but bloat can also occur on low forage/high concentrate diets due to a shortage of fibre.

When such predisposing diets are offered, the supplementary feeding of long forage will reduce the incidence but unfortunately not all cattle will eat sufficient amounts. The use of anti-foaming agents, such as peanut, linseed and paraffin oils administered as a drench, in the drinking water, applied to the flanks for licking or sprayed onto the pasture, will prevent the formation of the foam in the rumen.

Displaced abomasum

The displacement of the abomasum occurs around the time of calving and is becoming increasingly common in cows on high levels of feeding. The abomasum migrates within the abdominal cavity and often becomes trapped by the rumen, blocking the passage of digesta. In such cases appetite and milk yield are depressed and there is a loss of body weight which may result in ketosis.

A high forage ration in the dry period should ensure a low incidence of the problem. Treatment of the condition involves either turning the cow onto her back for a few minutes, or surgery to manually return the abomasum to its normal position.

Foreign bodies

It is not uncommon for metal objects such as wire and staples to be swallowed by the cow during feeding. These objects generally move into the reticulum, and these may penetrate the wall causing peritonitis which results in a raised temperature and a drop in appetite and milk yield. Further penetration may damage the pericardium of the heart and be fatal.

The only preventative measure is vigilance in ensuring that no metal objects pass into the forage or concentrate diet. Treatment normally involves surgery to remove the foreign body.

Udder problems

Mastitis

Mastitis is an inflammation of the udder tissue caused by one or more of a number of organisms which gain entry through the teat orifice.

The initial stages of infection are sub-clinical with no visual effect on the milk or udder. Some of these infections recover spontaneously, but many become clinical days or months after. Clinical mastitis is characterized by clots in the milk caused by the precipitation of milk proteins, by leucocytes and by epithelial cells. In severe cases the udder becomes swollen, the infection becomes systemic giving rise to a high temperature, and in some cases death occurs. Some cases of mastitis, particularly after calving, are not associated with infectious agents and are a result of the physiological changes in the udder following calving.

All stages of mastitis caused by infections affect milk yield and composition. Average depressions in the yield of infected quarters is about 15 per cent, but this varies greatly according to severity, and in extreme cases the quarter is so badly damaged that milk secretion is permanently stopped. The milk fat and lactose contents are also depressed to levels averaging 0.2 units less than for uninfected quarters.

Herd infection levels range from 10 to 90 per cent of cows and on average about 50 per cent of cows are infected, although less than 5 per cent will normally show clinical symptoms at any one time.

Over 90 per cent of infections are caused by the bacteria *Staphylococcus aureus, Streptococcus agalactiae, S. dysgalactiae* and *S. uberis*. The main source of these organisms is infected quarters and therefore most transmission of disease from cow to cow occurs during the milking routine. The preventative measures for controlling these infections are therefore concerned with good hygiene methods at milking including teat disinfection and routine antibiotic therapy at the time of drying-off.

Other important types of mastitis are contracted between milkings, the most important being those infections caused by *Escherichia coli* and *Corynebacterium pyogenes*. The *E. coli* type is increasing in importance and occurs mainly in early lactation in herds with low cell counts, housed in cubicles and fed on high-quality silage. The infections are picked up from dirty cubicle beds and other sources of contamination of the teat orifice. The result is normally a severe infection with a high temperature, loss of milk, and lost quarters and death are not uncommon. Control of the disease must have, as a main objective, keeping the udders clean. *Corynebacterium pyogenes* infections are mainly seen in dry cows and pregnant heifers in July and August (summer mastitis). The infection is thought to be transmitted by flies and invariably results in the loss of a quarter. The control of flies on and around the udder is thus the main control procedure, in conjunction with the routine dry cow antibiotic therapy.

Prevention of mastitis therefore involves:

1. Adoption of a hygienic milking routine including foremilking (which is a statutory requirement) to identify clinical cases for treatment with antibiotic, washing with clean water (preferably containing a disinfectant), drying with an individual paper towel, followed after milking by teat dipping/spraying with a recommended disinfectant (usually a solution of sodium hypochlorite, iodine or chlorhexidine).
2. Giving antibiotic as recommended by the veterinary surgeon to all cows at drying-off and to quarters with clinical mastitis.
3. Regular maintenance of the milking equipment.
4. Keeping the cows clean and dry.
5. In July/August using a long-acting insecticide spray or Stockholm tar on the udders of dry cows to prevent summer mastitis.

Any antibiotic treatment whether given by the intramammary or intramuscular methods should be associated with the discarding of all the milk for a prescribed period, which varies according to the antibiotic used. It is an offence to sell milk which includes an antibiotic.

Cell counts of milk give a general indication of the mastitis status of a herd. Non-infected quarters normally have counts of less than 100,000 cells/ml, but average herd levels are 400 000–500 000 cells/ml. The above recommendations for mastitis prevention should result in cell counts of less than 250 000 cells/ml.

Pendulous udders

The udder of the cow is likely to become more pendulous with age as the central suspensory ligament becomes weaker. This can result in the enforced culling of many high yielding cows.

The condition is heritable and therefore selection of sires which produce daughters with good udder conformation is important. Practices such as 'steaming-up' particularly with high protein diets before calving should be avoided as they lead to oedema and a stretching of the udder in the late dry period.

Foot problems

The incidence of problems associated with cows' feet has increased in recent years due to changes in cubicle housing and slurry systems, to silage feeding, to higher planes of nutrition and possibly to breeding. Such problems lead to the cow spending longer periods lying, to a reduced time spent feeding and consequently to depressed milk yields.

Where the problem is due to overgrown and misshapen hooves, foot trimming should be carried out. For most herds an annual session of trimming is adequate, and this will average about 30 per cent of the herd requiring treatment. Attention to the selection of sires that produce daughters with a low incidence of foot problems is an important longterm strategy.

Infections of the foot are caused by organisms penetrating the tissue of the hoof. A high incidence can occur when the hooves are soft and when cut or bruised. Predisposing conditions are: wet, muddy gateways and cow tracks in the summer which are further aggravated by stones and in the winter cows standing in slurry, and walking on rough concrete. The best means of prevention are to keep the feet as dry as possible, to avoid having rough concrete and stony tracks, and to put the cows through a weekly footbath of 5 per cent formaldehyde or 10 per cent copper sulphate solution which not only hardens the hooves but also kills organisms on the surface.

Many foot problems are associated with feeding (laminitis) and this can occur in all four feet. It is seen in cows in early lactation on high planes of nutrition particularly on diets containing a high proportion of non-protein nitrogen and/or starchy concentrate. It is characterized by a swelling above the hoof which is warm to the touch, by a softness of the hoof and a predisposition to penetrations. To prevent the disease, overfeeding of these dietary components should be avoided. If large amounts of concentrates are fed, a little-and-often feeding system is beneficial as this leads to a more stable pH in the rumen and to a reduced chance of laminitis problems. The protein content of the concentrates should also be selected to ensure the crude protein content of the total DM does not exceed 18 per cent.

Other diseases

Brucellosis

In the past, brucellosis caused by the organism *Brucella abortus* was a considerable problem in herds. It is a contagious disease causing abortion and infertility in cattle, and can be contracted by man producing the condition known as undulant fever.

Initially the disease was prevented by vaccinating calves

with a low-virulence strain of the organism S 19, but this procedure was superseded by the national eradication scheme which has been successful in its objective of making the UK brucellosis-free.

Tuberculosis

Until the 1950s, tuberculosis was the most widespread disease of dairy cattle, and it could be transmitted to man in milk. The national eradication scheme has since resulted in all herds being tuberculin-free.

Bovine leucosis

This includes a group of diseases characterized by growth of the lymphocyte forming tissues. The most common type, Enzootic Bovine Leucosis (EBL) causes tumour formation mainly in cows 4–8 years of age. The pattern of the disease suggests there is vertical transmission from one generation to the next, but horizontal transmission of the disease also occurs. A lymphocyte count is a useful method for screening herds for EBL but is unreliable for individual cows as about 10 per cent of cases have a normal count.

The greatest risk was originally from imported cattle but such animals are now tested. The clinical disease has, however, an extremely low incidence in the UK.

Parasites

Adult cattle are normally resistant to lungworm (*Dictyocaulus viviparus*) and to parasitic gastroenteritis caused by stomach worms such as *Ostertagia ostertagi*, provided they have grazed contaminated pasture during the rearing period. In some herds, however, a response in milk yield to anthelmintic treatment at calving has been shown.

Liver fluke can also affect the performance of dairy cows grazed in wet, boggy areas. The disease can be controlled by routine drenching or injection with anthelmintic, but in the long term, drainage of wet areas of land is the best solution.

Warble flies cause considerable damage to the hides of cows when the grubs emerge on the cows' backs in the spring. Control of the disease is brought about by applying a liquid systemic treatment along the back of the cow in October/November, or by treating clinical cases in a similar manner in spring. An injection is also available which not only kills warbles, but also controls lungworms, stomach worms and lice.

Milk management

8

The trend towards larger dairy herds which are looked after and milked by fewer staff has meant that quick and efficient milking has become an increasing priority. Efficient milking is concerned with the production of milk under hygienic conditions by cows and staff who are not under stress.

The process of lactation involves the secretion of milk by the alveolar cells of the mammary gland, its storage in the alveoli and ducts, milk ejection under the control of the hormone oxytocin, and finally its removal from the udder. These aspects are covered in the sister Handbook in this series, *Lactation of the dairy cow* by Colin T. Whittemore.

Extraction

The basic principles of milk extraction have remained essentially unchanged since the beginning of the century. A vacuum is applied within the teat-cup liner, to the teat end and the resulting pressure difference between the teat sinus and the open liner results in the opening of the teat sphincter and milk flows out from the streak canal. Applying a continous vacuum to the teat end can result in damage to the teat, and therefore pulsation is used with alternate opening and closing of the liner (Fig. 8.1).

Increasing the vacuum level increases the flow rate of milk from the teat, but this results in an increased amount of residual milk remaining in the ducts and can also lead to

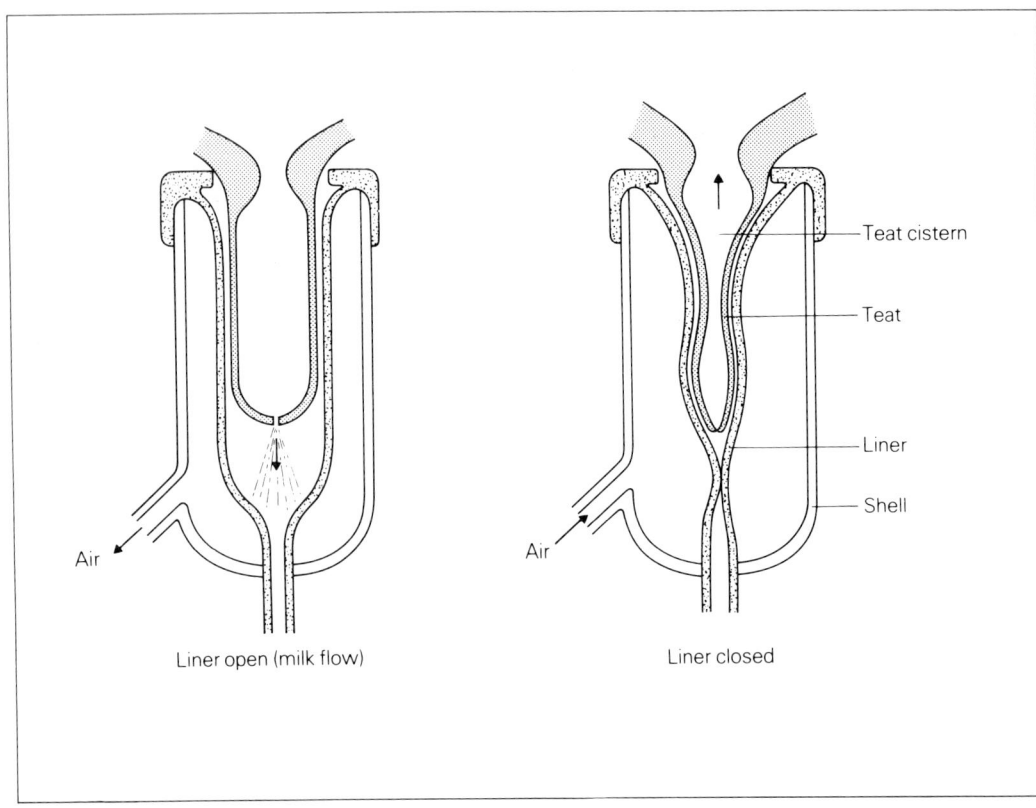

Liner open (milk flow)   Liner closed

Teat cistern

Teat

Liner

Shell

Air

Air

Figure 8.1 Action of a teat-cup liner

teat damage. Most milking machines work at a vacuum level of 45–50 kPa (13–15 inches of mercury).

The flow rate of milk is also increased by increasing the pulsation rate (pulsation cycles per minute) and by widening the pulsation ratio (proportion of each pulsation cycle that the liner is more than half open). The recommended rates arc 45–70 cycles per minute at ratios of 50–75 per cent.

A range of rubber and synthetic rubber teat-cup liners are available. The characteristics of a liner are that they should give a low incidence of teat slip, i.e. they should remain held on the same part of the teat throughout milking, and they should not damage the teat orifice. The claw-pieces which collect the milk from the four teat cup liners, are usually fitted with air-bleeds which are small holes aiding the speedy removal of milk from the liner and into the long milk tubes.

Flow rate

Milk flow normally commences within 10 seconds of applying the cluster, it then builds up within 30 seconds to a steady rate (peak flow) which is mainly controlled by the dimensions of the teat orifice and the streak canal. Milk flow rate then declines as the teat sinuses are emptied, and if automatic cluster removal (ACR) is used the cluster is taken off when the flow rate falls to less than 0.5 kg/per minute (Fig. 8.2).

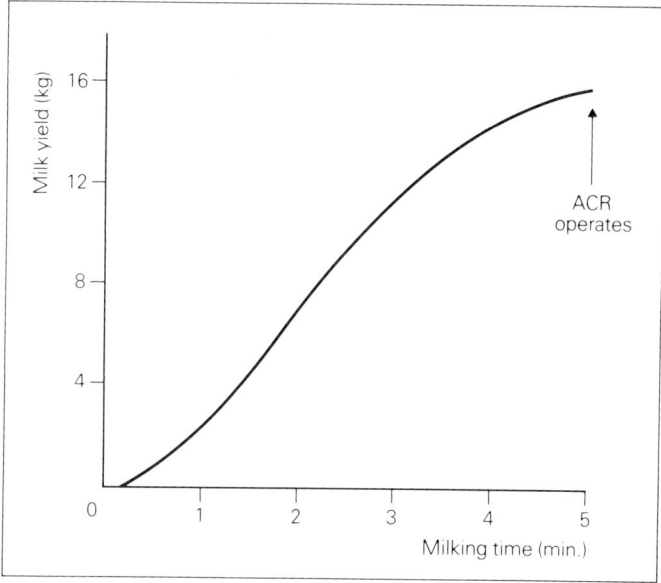

Figure 8.2 Milk flow during the milking of a cow

There is a considerable variation between cows in milk flow rate (Fig. 8.3). In modern high-throughput milking parlours, slow-milking cows tend to cause undue delays and often have to be culled from the herd. As the characteristics of the teat are highly heritable, this selection procedure has long-term benefits.

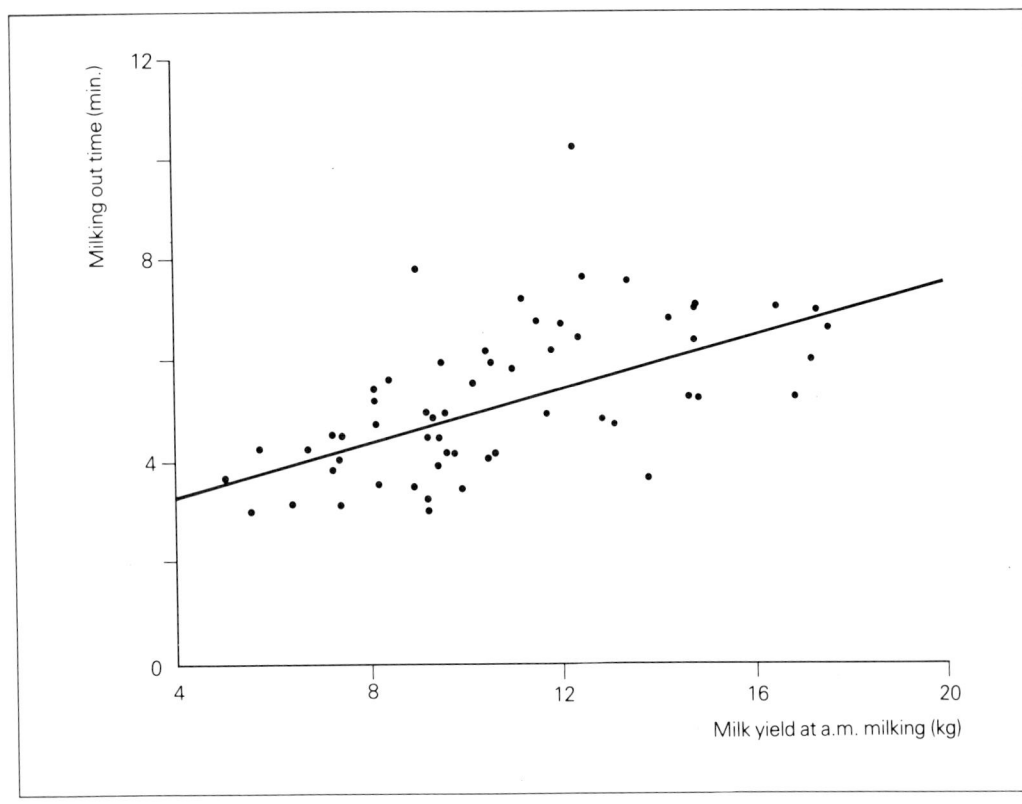

Figure 8.3   The milking out times of individual cows

Milking interval

Milk secretion is a continous process in the udder and the rate of secretion may be constant for 15 hours or more. Thereafter the increasing intramammary pressure reduces the rate of secretion. For the majority of cows there is little evidence that unequal milking intervals of twice-daily milking, e.g. 14 hours and 10 hours or even 16 hours and 8 hours, give lower yields than milking at 12 hour intervals. This has been shown with cows yielding over 6 000 kg milk per lactation. The milk fat content is always greater after the shorter interval than after the longer interval when milking times are unequally spaced.

In herds with yields over 6 500 kg there may be merit in moving more towards 12 hour intervals to reduce intramammary pressure, and also to consider three times per day milking at approximately 8 hour intervals. This latter practice normally increases milk yields by 10–15 per cent at all stages of lactation, which suggests that its effect is not simply the reduction in intramammary pressure. The additional stimulation of the third milking appears to produce an additional hormone response which gives rise to a higher milk secretion rate. The milk fat content is normally slightly reduced by milking three times per day. Two additional benefits of this practice are that less damage is made to the udder as it is not as distended in early lactation, potentially leading to a longer herd life, and it may also reduce the incidence of mastitis.

Performance

Milking takes place at each end of the working day, and therefore the duration of milking tends to determine the length of the working day. Milking performance is thus an important factor not only in labour costs but also in job satisfaction.

As the number of cows for which each man is responsible is increasing, the milking performance (number of cows milked per man hour) has to receive increasing attention. The milking period should have two main objectives – efficient milking and the detection and treatment of clinical mastitis. Other stockmanship duties are much better performed between milkings.

The milking performance is determined: (a) by the work routine per cow; and (b) by the waiting time by the operator for cows to milk out.

The work routine represents the time required in minutes per cow for the operator to perform his milking duties on that cow, and is the major factor affecting milking performance. Examples of work routines are shown in Table 8.1. The work routine can be reduced and the milking performance increased by omitting parts of the routine (providing there are enough milking units to keep the operator occupied, otherwise waiting time will occur).

It is a necessary requirement that the milking machine is applied to clean teats, and washing teats followed by drying is important for the production of clean milk. The

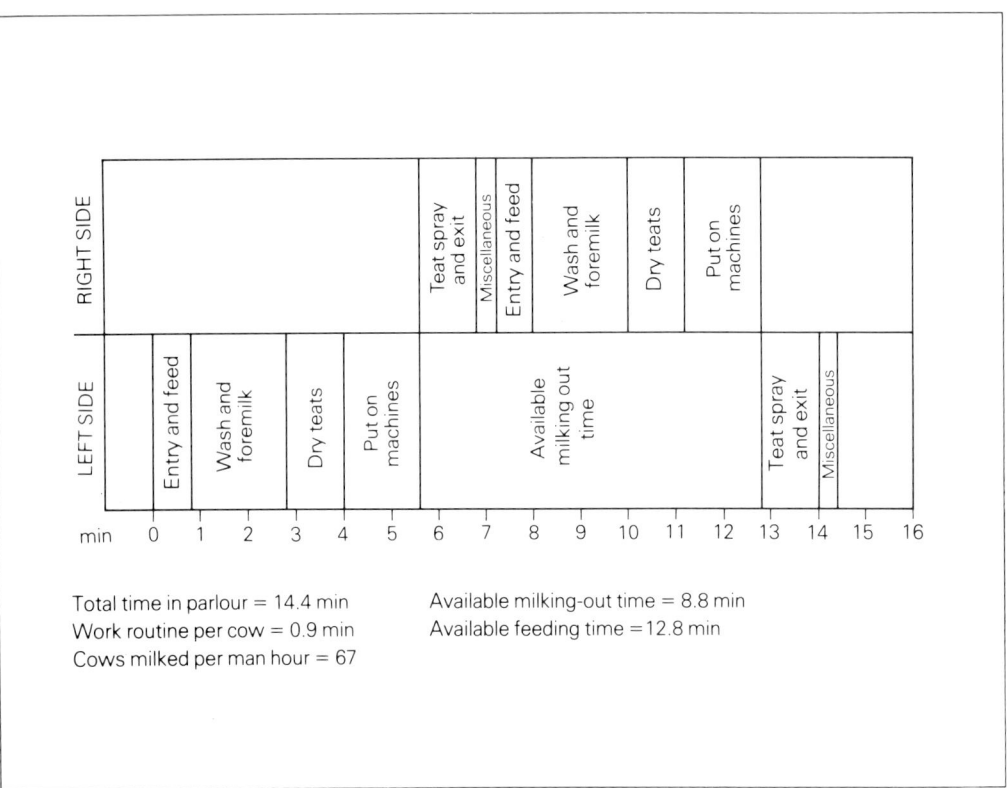

Left side labels: LEFT SIDE — Entry and feed, Wash and foremilk, Dry teats, Put on machines, Available milking out time, Teat spray and exit, Miscellaneous

Right side labels: RIGHT SIDE — Teat spray and exit, Miscellaneous, Entry and feed, Wash and foremilk, Dry teats, Put on machines

min 0 1 2 3 4 5 6 7 8 9 10 11 12 13 14 15 16

Total time in parlour = 14.4 min
Work routine per cow = 0.9 min
Cows milked per man hour = 67

Available milking-out time = 8.8 min
Available feeding time = 12.8 min

Figure 8.4 A diagram of the work routine in a single operator 16 stall by 16 unit herringbone parlour

Table 8.1 Work routines in a herringbone parlour

	A (full routine)	B (with ACR)	C (reduced routine)
Entry and feed	0.10	0.10	0.10
Wash teats	0.15	0.15	—
Foremilk	0.10	0.10	0.10
Dry teats	0.15	0.15	—
Put on machine	0.20	0.20	0.20
Take off machine	0.10	—	—
Teat dip/spray and exit	0.15	0.15	0.15
Miscellaneous	0.05	0.05	0.05
Total	1.00	0.90	0.60
Maximum performance* (cows/man hour)	60	67	100

* Assuming there is no waiting time

Table 8.2 Milking performance in relation to work routine and waiting time

Work routine (min./cow)	1.00	1.00	0.60	0.60
Maximum performance (cows/man hour)	60	60	100	100
Waiting time (min./cow)	—	0.10	—	0.10
Total time per cow (min.)	1.00	1.10	0.60	0.70
Actual performance (cows/man hour)	60	55	100	86

development of ACR has taken from the routine the element of cluster removal and has also removed the bad

habit of machine stripping (applying weight to the claw-piece by hand towards the end of milking), which has little beneficial effect on milk yield. Teat dipping or spraying with disinfectant after removal of the cluster is beneficial in mastitis prevention. The miscellaneous time includes the time spent washing clusters and floor during milking, carrying out mastitis treatments, attending to clusters which fall off, etc. A flowchart of a work routine in a 16 stall × 16 unit parlour is shown in Fig. 8.4.

The waiting time is the average time per cow spent waiting for the cluster to be removed. This situation arises with high yielding cows, with slow milking cows and where the milking parlour has too few milking units. The average waiting time per cow is added to the average work routine per cow to give the average milking performance level. The example in Table 8.2 illustrates how waiting time can have a significant effect on milking performance, particularly where short work routines are practised.

Milking systems

Cowshed (byre) milking

Many small herds (under 60 cows) are milked in cowsheds although the numbers are declining. Most of these systems involve milking directly into an overhead glass or stainless steel pipeline and thence to the bulk tank. The pipeline removes the chore of carrying milk to the dairy as with bucket milking systems.

One operator can handle 4–6 units and if a full work routine is practised, a milking performance of up to 60 cows/man hour is possible, but in most herds, due to waiting time and to machine stripping, the rate is rarely more than 30 cows/man hour.

Abreast parlour

This is the simplest type of milking parlour and was originally developed as a movable bail. The most common layout is the 8 stall × 4 unit parlour (Fig. 8.5). The cows stand side by side in stalls which have a front gate and a chain at the rear. The floor level of the stalls is often raised to 250–400 mm to allow an easier working height for the operator.

Herringbone parlour

The fixed herringbone parlour is the most common type in use (Fig. 8.5). The cows stand at an angle of about 30 ° on both sides of a 75–85 mm deep pit. There are two main types, the 1 stall and the 2 stall per unit layouts. In the latter case the recorder jars are situated in the centre of the pit and each milking unit alternately milks cows on the left- and right-hand sides. These two layouts have the same potential throughput but the 1 stall per unit type has a longer available milking out time, so if a short parlour is employed (10 × 10 or 12 × 12) it will have less waiting time and therefore a slightly higher performance than the 2 stall per unit type (10 × 5 or 12 × 6).

Another advantage of 1 stall per unit layouts is that the pit is uncluttered and therefore more pleasant to work in. The main disadvantage of this layout is its extra cost as double the amount of equipment is required compared with the 2 stall per unit.

Rotary parlours

The equipment and cows in rotary parlours are carried on a rotating platform. Three main types are available, the rotary abreast, rotary tandem and rotary herringbone (Fig. 8.6).

In the rotary tandem the cows stand nose to tail with the operator working inside the circle. This layout requires a large building to house the parlour and is therefore expensive. A number of small 8 stall parlours are in use, but high yielding cows have to travel round the parlour more than once before being milked out, and the performance is no better than for fixed herringbones.

The rotary abreast is the simplest and most common type. The main advantage of this layout is its compactness and therefore it does not require a large building. The

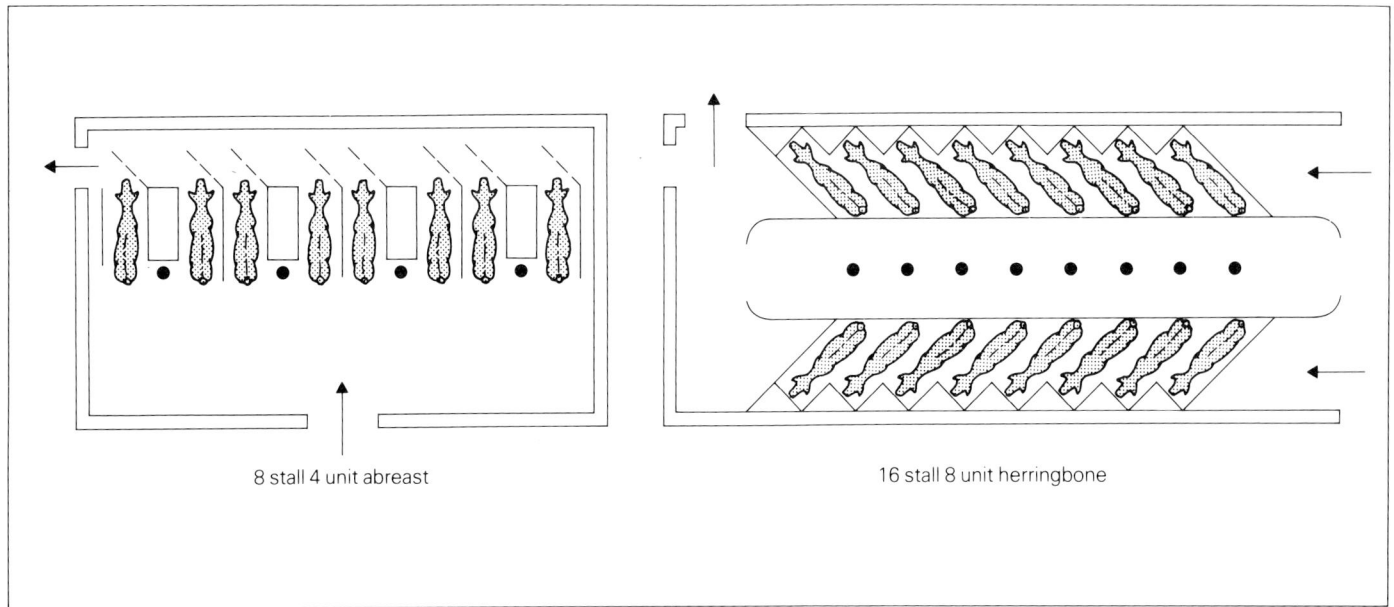

Figure 8.5 Fixed abreast and herringbone parlours

disadvantage of the outer operation type where the cows face inwards is that the operator cannot see the machines and cows on the far side. The inner operation type where the cows face outwards overcomes this problem and appears to combine the best points of rotary milking.

The rotary herringbone is the most complicated parlour and thus the most expensive. The cows stand in herringbone fashion facing outwards.

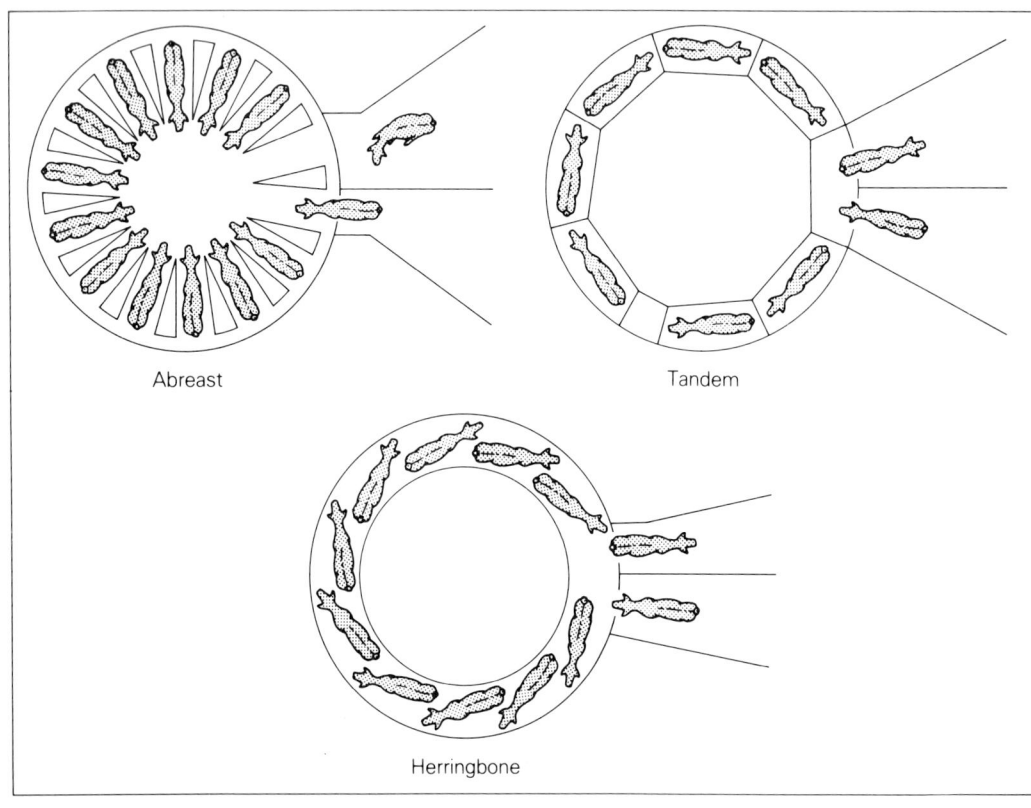

Abreast

Tandem

Herringbone

Figure 8.6 Three types of rotary parlours

Rotary parlours are appropriate for herds of over 200 cows where very high throughputs of over 100 cows per man hour are required. The rotary movement is ideal for automation in cow entry and exit, feeding and teat spraying, thus allowing a reduction in the work routine per cow and resulting in a high level of performance. The main disadvantage of rotaries compared with fixed herringbones is the higher maintenance cost and lower reliability due to having moving parts.

Choice of parlour

The factors which should be taken into account in installing a new milking parlour are as follows.
1. The required milking performance. The number of cows required to be milked per hour should first be specified.
2. The proposed work routine. The choice of work routine will depend on what degree of control is required in hygiene and mastitis prevention, and how much automation is to be introduced. A short work routine, such as 0.7 minutes/cow will require a longer herringbone parlour than where a long work routine of 1.0 minutes/cow is practised in order to prevent waiting time, i.e. a short routine gives a shorter available milking out time.
3. The future yield and calving pattern. The yield level and calving pattern will determine the milking out times of the cows at different times of the year, and therefore dictate how long the parlour should be in order to prevent waiting time.
4. Capital availability. Milking parlours are expensive, and different choices are available at the same price. In general if a fixed herringbone is the choice it is better to invest in a long 2 stall per unit layout than a short 1 stall per unit at a similar price, such as a 16×8 rather than a 12×12. If necessary, conversion to a 1 stall per unit system can be made at a later date without additional building costs.

For herds of under 80 cows, cowshed or abreast parlours will give an adequate performance, but for larger herds of up to 200 cows fixed herringbones with one man operation are the most satisfactory type of installation. For herds of over 200 cows, rotary parlours should be considered as an alternative to long, 2-man herringbones (24–32 stalls) or to 2 separate one man herringbones.

Milking parlour equipment

There are many different types of milking equipment and layout available, but a typical example is shown in Fig. 8.7. Milking can be direct to pipeline, through static flow-meters to pipelines, or into recorder jars set at low level, eye level or high level.

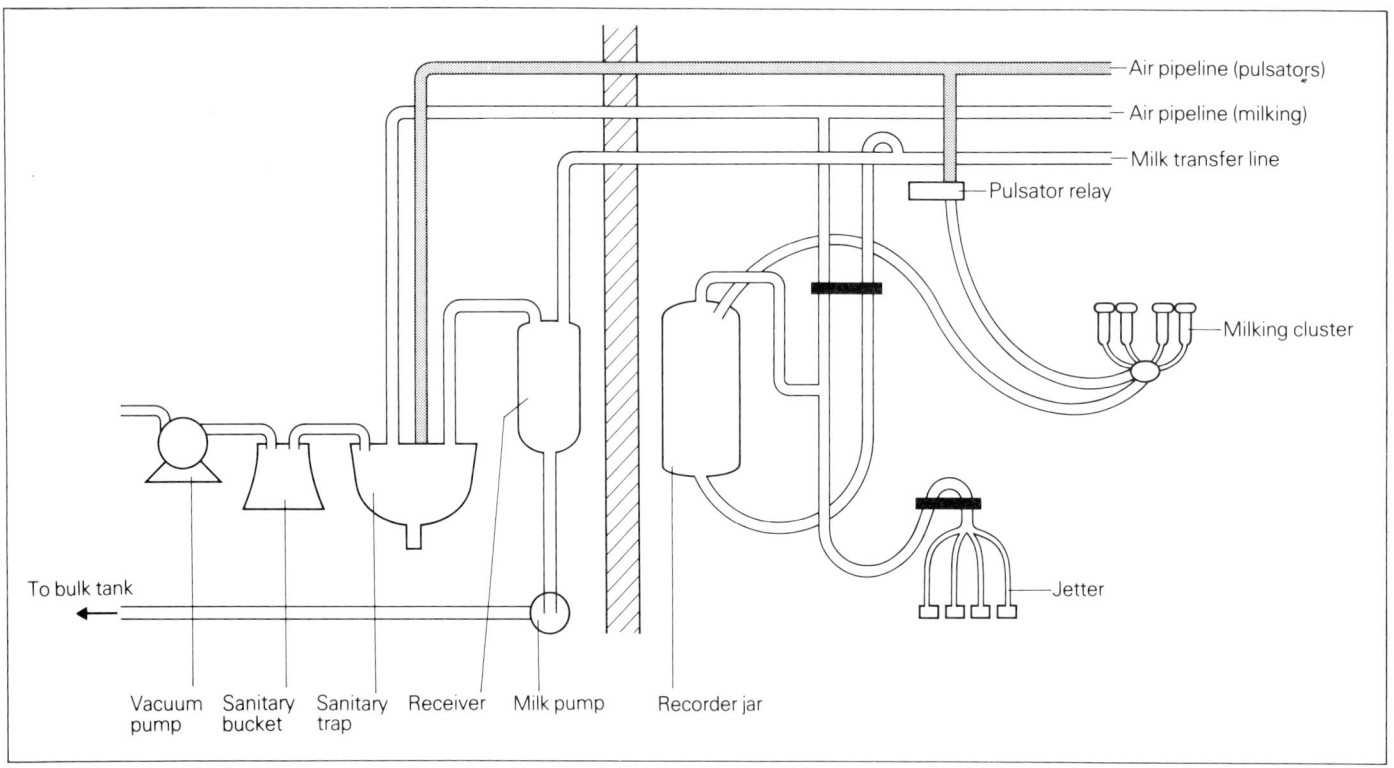

Figure 8.7 The milking equipment for a milking parlour with recorder jars and using circulation cleaning

Production of clean milk

The hygienic production of milk should be a high priority for the dairy farmer, if in the long term he wishes there to be a strong market for milk and its products.

Contamination of milk is caused by dirt and bacteria, the former from unclean teats, and the latter both from the exterior and the interior (infections) of teats, and from the milking equipment. Bacterial counts below 100 000/ml are normally required by the Milk Boards, but counts of below 10 000/ml should be the objective of all producers.

Udder cleanliness

The first prerequisite for clean milk is that cows are housed under clean, dry conditions, and if grazed that they are not walking through deep mud. Soiling with dung, bedding materials or mud between milkings can result in a large sedimentation and high bacterial counts in the milk.

Although washing of teats reduces the level of sediment it will have little effect on mastitis prevention unless drying is also carried out. The use of a disinfectant in the washing water has a dual purpose – it is beneficial in killing some bacteria on the teat surface, and in hose washing-systems it also prevents the build-up of bacterial contamination in the washing water which can give rise to udder infections.

Cleaning of equipment

There are two main systems of cleaning pipelines and milking equipment in cowsheds and parlours – circulation cleaning and acidified boiling water (ABW) cleaning. When the milking is completed the milk line is drained of all milk and the filter and pipe leading to the bulk tank are removed. The external surfaces of the milk clusters are brushed and the teat cups attached to the jetters. The plant is then set to the wash position.

Circulation cleaning is the most commonly used method and is normally a three-stage process, with a pre-rinse of warm water, a recirculated hot wash with detergent solution, followed by cold-water rinse. Although the system uses a chemical disinfectant it depends very much on the water temperature (initially 85 °C) for its effect. The process is completed in about 15 minutes.

The ABW method relies mainly on heat to clean and disinfect the system. Water at almost boiling point is passed through the system without a pre-rinse and discharged at the end of the circuit. A temperature of at least 77 °C is required on all surfaces for at least 2 minutes for the method to be successful, and acid is added in the first half of the wash to assist cleaning and prevent deposits forming. The method requires up to a third more water at a temperature 10 °C higher than for circulation cleaning, but the process only takes about 6 minutes.

The cleaning of bulk tanks is normally a three-stage process with a cold-water rinse, a cold or warm spray with disinfectant, followed by a cold rinse. The system can be automated and is started by the tanker driver as soon as he has finished pumping out the milk. However, at regular intervals additional brushing with hot water and disinfectant may be necessary to prevent the build-up of deposits on the internal surfaces of bulk tanks.

Buildings, equipment and labour

9

The fixed costs (overheads) of the milk production enterprise have assumed increasing importance as herd sizes have increased, as systems have become less labour intensive, and as interest charges have increased. The balancing of costs between investments in buildings and equipment, and in labour use, is an essential aspect in efficient management. Such decisions can have long-term effects on the profitability of the enterprise.

Objectives for dairy buildings and equipment

When new buildings are erected, or when present buildings are adapted for a change of system, the objectives should be to:

1. Protect the staff and cattle from inclement weather.
2. Provide a comfortable lying and lounging area for cows which gives a low incidence of injuries.
3. Provide adequate feeding and water facilities.
4. Provide facilities for calving and handling cows.
5. Provide an efficient milking system and dairy.
6. Provide ample room for storage of feeds, and arrange for efficient food handling.
7. Provide a satisfactory system for the storage and disposal of effluent.
8. Satisfy all legal requirements for planning and building regulations, and for milk and dairies regulations.

The dairy building

There has been a trend towards the erection of more specialized buildings for milk production systems which are less adaptable for conversion for use in other enterprises. This has led to buildings being designed for functional purposes which, even though legal requirements are satisfied, can lead to neglect of aesthetic considerations. Where possible a new building or an extension to a present building must be considered in relation to the whole farm system. The planning policy should be long-term; too often a new building is erected in a position which prevents further expansion in the future.

Construction

The two main types of building construction are:
(a) those with concrete, steel or timber frames, with walls of concrete blocks and space-boarding, and roofs of corrugated asbestos cement, steel or aluminium sheets; and
(b) prefabricated timber kennel buildings in which extensions of the cubicle division uprights act as supports for the sheeted roof.

The advantages of framed buildings (Fig. 9.1) are that they have a long life-span, have low maintenance costs and are more adaptable for use in other enterprises if the milk production enterprise is terminated. The kennel type building has a lower initial cost, tends to be warmer and less draughty, but has a higher annual maintenance cost and shorter life.

Layout

As herd sizes have increased, the proportion of cows housed in loose-housing systems compared with cowsheds has increased due to the reduction in labour requirements.

There is a wide range of possible building layouts to facilitate the housing, feeding, and milking of dairy cows and the storage and disposal of slurry. It is therefore unusual to see even two layouts which are exactly the same. The important priority is to satisfy the objectives set out earlier.

Consideration has to be given to the interaction of the basic components (Fig. 9.2). The housing area (cubicles or straw yard) should be serviced by a satisfactory feeding system and effluent removal and storage system. There should be a calving area adjacent, which is designed to allow efficient (mechanized where possible) feeding and bedding removal. The collecting area and milking parlour should be sited so that there is an efficient flow of cows to and from the parlour. The facility to clean the parlour and collecting area into the effluent storage system is also necessary. An example layout is shown in Fig. 9.3.

Figure 9.1 Steel span and timber kennel buildings for dairy cattle

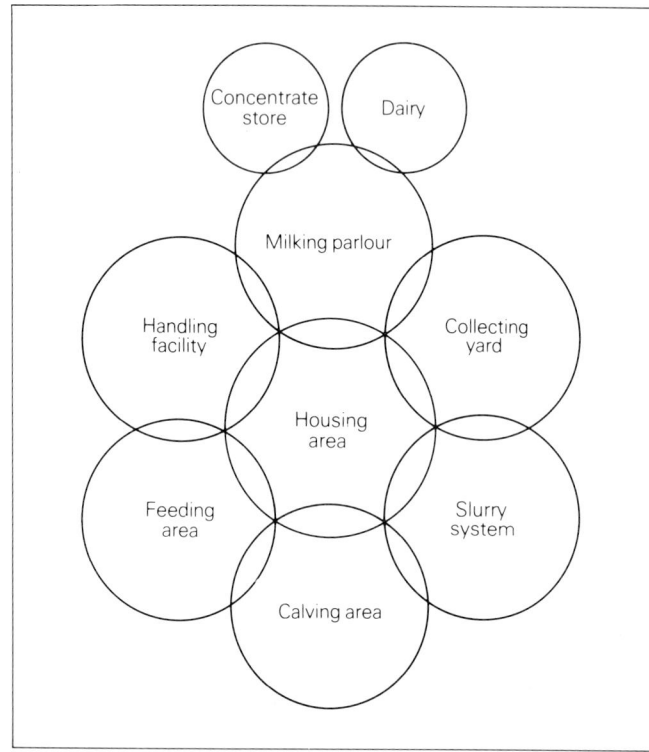

Figure 9.2   The components of a dairy unit

Housing area

Traditionally cows were housed and milked in cowsheds but the trend, due to increasing herd sizes and to the use of silage as the main forage, is towards separating the housing and milking areas. Loose-housing gives increased cow comfort, it allows the cows to have exercise, and most importantly the feeding, bedding and cleaning can be mechanized thus leading to a low labour requirement.

Strawyards

In arable areas where straw is a relatively low-cost commodity, loose-housing in strawyards is a commonly used system (Fig. 9.4). One advantage compared with cubicle housing is that the manure is handled as a solid, and when mechanically removed from the strawyard it can be stored almost anywhere on the farm until the time for spreading.

Many straw yards are based on a bedded area of 5–6 m^2/cow, but if the system is associated with a scraped concrete floor area, such as with self-feed silage, or with a feeding passage, the bedded area can be reduced to 3–4 m^2/cow. Total straw usage over a 6-month winter will range from 0.5 to over 1.0 t/cow depending on the system. The floor of the strawyard can be of earth, chalk or concrete.

Entry and exit from the strawyard should be as wide as possible, otherwise trampling of the bedding will occur

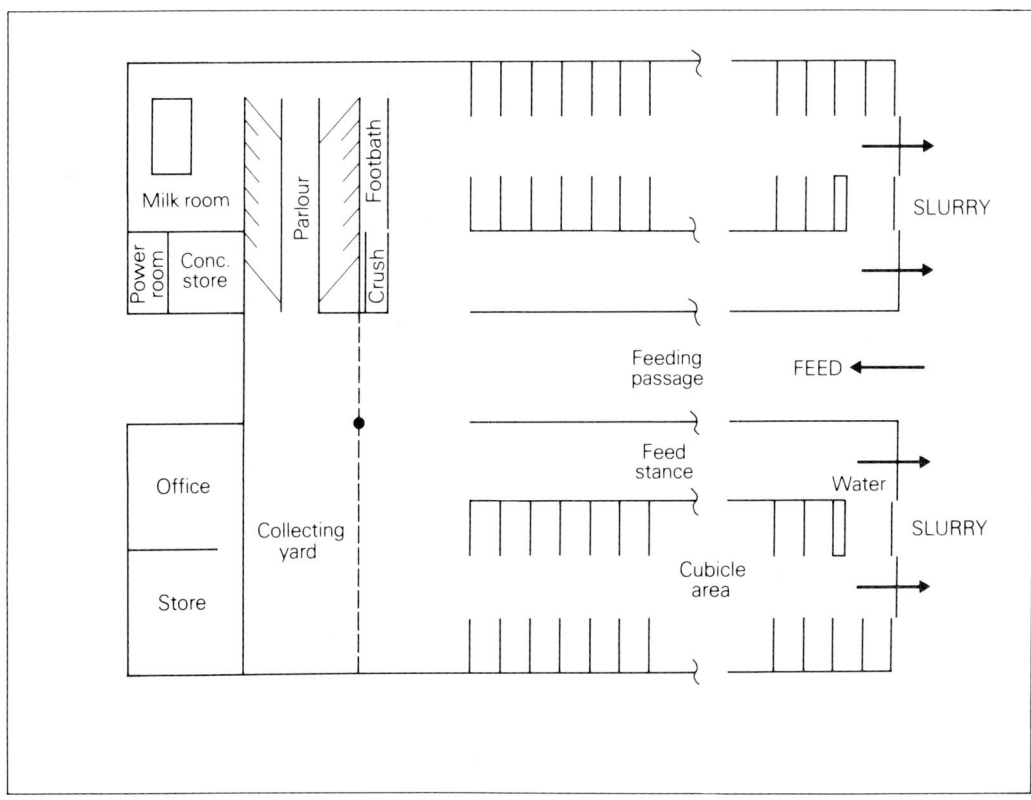

Figure 9.3 An example layout of a dairy building

Figure 9.4 A strawyard system of housing

leading to wastage and dirty cows. The feeding and watering facilities should be sited on concrete areas adjacent to the bedded area, as the height of the bedding is continually rising, and as wastage from water bowls or

troughs can lead to a greater than necessary straw usage. The concrete area should be at least 2 m^2/cow.

Cubicles and kennels

The cubicle system of cow housing was developed in the early 1960s and this subsequently led to the development of low cost kennel systems. The attractions of the cubicle system are the economy of bedding use which is particularly important in non-arable areas where straw is expensive. However, the capital cost of concrete, cubicle divisions and the provision of a satisfactory slurry removal, storage and spreading system, can be very high.

The dimensions of the cubicle are more important than the design of the cubicle division. Suggested minimum sizes are shown in Table 9.1. The length measurement is much

Table 9.1 Dimensions of cow cubicles

Liveweight (kg)	Length (m)	Width (m)
Over 650	2.20	1.20
600–650	2.10	1.15
550–600	2.05	1.10
500–550	2.00	1.05
450–500	1.95	1.00

more critical than the width, and a longer than average cubicle with a correctly placed head-rail is the ideal.

The cubicle bed is normally of concrete although bitumastic and rammed chalk are alternatives. The advantage of bitumastic is that it has better insulating properties than concrete, although concrete beds should be insulated with a layer of drainage tiles or polystyrene beneath the surface. The problems with rammed chalk are that it does not retain its profile and tends to be 'hollowed out' with cows lying and standing.

A heelstone or kerb may be used to retain bedding in the cubicle (Fig. 9.5), but many farmers prefer to have a sloping bed with no heelstone, so that any dung or urine which falls on the bed can run off or be easily scraped off. However, if the head-rail is correctly placed (450–500 mm from the front of the cubicle) very little dung and urine should fall on the cubicle bed.

The bedding materials used for cubicles depend on price, availability and the suitability for the slurry storage system. The four main types of material are as follows:

Sawdust and **wood shavings** are widely used but are often difficult to obtain and expensive. It is essential that the material is dry otherwise a build-up of coliform organisms can occur during storage, giving rise to coliform mastitis when placed in the cubicles. The winter usage ranges from 100 to 300 kg/cubicle.

Figure 9.5   A cubicle system of housing

Straw is commonly used on farms with an arable enterprise. The material should be chopped to reduce the amount carried off the bed on the cows' feet. The winter usage ranges from 200 to 400 kg/cubicle.

Sand is often used where there have been coliform mastitis problems in herds. The material is inert and therefore not a good medium for the growth of bacterial populations. Unfortunately the application of sand to the cubicle beds is more laborious than for other materials, and can be very abrasive on slurry machinery. Also it should not be used where the slurry is stored in a tank as it tends to settle to the bottom. The weight of sand used per cubicle depends very much on its moisture content, but will range from 500–900 kg/winter.

Cow mats, which are generally made of rubber, are an expensive item, but do give increased insulation and comfort to the cow. Normally they require some bedding material to be applied to absorb any slurry carried onto the bed by the cows' feet.

Other bedding materials, including lime and shredded paper, are used on a small number of farms.

The passageways between cubicles range in width from 2.0 to 2.5 m depending on the cleaning mechanisms – narrower passageways are commonly seen if there are concrete slats or automatic scrapers.

Feeding facilities

The amount of feeding facilities on a dairy farm depends on the size of farm, the availability of labour and whether the desired objective is towards individual cow, or to simplified group, feeding.

Forage storage

Silage is mainly stored in bunker (clamp) silos and pits, although a small amount is stored in tower silos. The advantage of bunker silos and pits is their low capital cost compared with towers although their dry matter losses in ensiling are generally greater (10–25 per cent compared with 2–10 per cent). Also grass can be ensiled in bunkers and pits with a suitable additive at low DMs, whereas for safety reasons grass for tower silos should be ensiled at over 30 per cent DM. The longer wilting time required for this material encourages farmers with tower silos to delay the cutting date to when the grass is more mature, in order to reduce wilting time. The result is a lower digestibility forage.

Bunker silos may be of concrete, concrete blocks or panels, or timber (Fig. 9.6). The roofing of silos is unnecessary where the silage is mechanically removed for feeding, if the silage is well sealed with a polythene sheet. However, with self-feeding most farmers prefer to protect the cattle and staff from the weather by roofing the silo area. The space requirements for silage range from 1.1 to 1.7 m^3/t of fresh material depending on the DM and the

Figure 9.6 Different types of silo for ensiling grass

Figure 9.7 Concentrate store with auger delivery to the milking parlour

degree of compression (rolling) which is carried out. The concrete base of the silo should slope away from the emptying face, where self-feeding is practised, to prevent urine from entering the silo, and where mechanical emptying is practised in unroofed silos to prevent rainwater from entering the silage.

Concentrate storage

There is an increasing trend towards bulk storage and handling of concentrate feed. For concentrates fed in the parlour this is either stored in a loft above the parlour or in a metal or timber bin outside the building, from where it is conveyed to the feed hoppers (Fig. 9.7). If bagged concentrates have to be used, increasingly they are stored and conveyed on pallets, by means of a fork-lift attachment.

Milking facilities

The demise of cowshed milking has meant that new milking installations are now concerned with milking parlours of one type or another. These are specialist milking sheds aimed at a high throughput of cows, and require a collecting yard, parlour, exit race, milk room, power room and in most cases a concentrate store (Fig. 9.3, p. 135).

The production and handling of milk is the subject of legislation which has the objective of safeguarding the consumer against disease and poor compositional standards. These regulations, generally known as Milk and Dairies Regulations, are covered by different Acts for England and Wales, Scotland and Northern Ireland.

Collecting yard

The throughput of cows is very dependent on a well-designed collecting yard. A space allowance of 1.4 m^2/cow should be allowed.

Rectangular collecting yards should have the entrance at the far end from the milking parlour, they should be long and narrow, and have a funnel arrangement at the entrance to the parlour. Backing gates or electrified wires to push the cows forwards can be incorporated which are operated by ropes from the parlour. In circular collecting yards, backing gates are often easier to operate using a variety of mechanisms, including water pressure and the winching of a cable attached to the gate.

To prevent slurry from entering the parlour it is preferable to have a small step into the parlour and the collecting yard should slope away from the parlour (approximately 1 : 30). All the washings from the collecting yard should be returned to the slurry storage system.

Milking parlour

The type of milking parlour is the main determinant of the size and shape of the building required in a new installation. Conversely if an existing building is used, this may dictate the type of parlour to be installed.

In many parts of Britain, doors at the entrance and exit of milking parlours are needed to comply with the regulations. It is important that these have adequate width (850–950 mm). Much of the milking is carried out in the hours of darkness and ample fluorescent lighting in the pit is essential. For the comfort of the operator an insulated ceiling (in roofed parlours) helps in conserving heat, and some form of overhead electrical heating is normally installed in the pit.

The parlour walls are normally constructed of brick or blockwork, but the internal finish has to be smooth and impermeable to comply with the regulations, and to assist in cleaning. Most walls are cement-rendered and treated with a paint which will give a hard-wearing and long-lasting finish, such as a rubberized paint. In some parlours the walls are tiled, but these are rarely successful as water tends to get behind the tiles and subsequently they crack or drop off. A more satisfactory development is the use of glazed blocks which provide the wall and smooth finish in one.

Milk room

The room must be sited for easy access by the operator from the milking parlour and by the milk tanker for the collection of milk.

The size of the room depends mainly on the space occupied by the bulk tank(s), but 750 mm clear space must be allowed around each bulk tank. Other equipment which is normally sited in the milk room includes the releaser jar and milk pump, the wash troughs for circulation cleaning, the bulk tank cleaner and possibly the water heater.

For most bulk tanks, a ceiling height of 2.4 m is adequate and due to the amount of washings going on to the floor, the latter should have a slope of about 1 : 40 towards the external drain gulley. The walls should be tiled or rendered to a height of 1.5 m.

Milk is stored in bulk tanks at a temperature which should not exceed 4.4 °C. At higher temperatures the growth of psychotrophic bacteria is rapid. The objective in bulk tank cooling is that the milk achieves the desired temperature within 30 minutes of the end of milking.

Most refrigerated bulk tanks in Britain are chilled water cooled with an ice bank contained within the structure of the tank (Fig. 9.8). The compressor, condenser and receiver (condensing unit) are normally sited in the power room along with the vacuum pump for the milking equipment, due to the amount of heat generated from this unit.

In order to conserve the heat produced by the condensing unit, many farms install heat recovery units which are connected between the refrigeration compressor and the air-cooled condenser. Hot gas from the compressor condenses in passing through the heating coil of the heat recovery unit giving up its heat to the water. The heated water can then contribute to the circulation cleaning or washing-up water, be used for udder washing, or as drinking water for the cows.

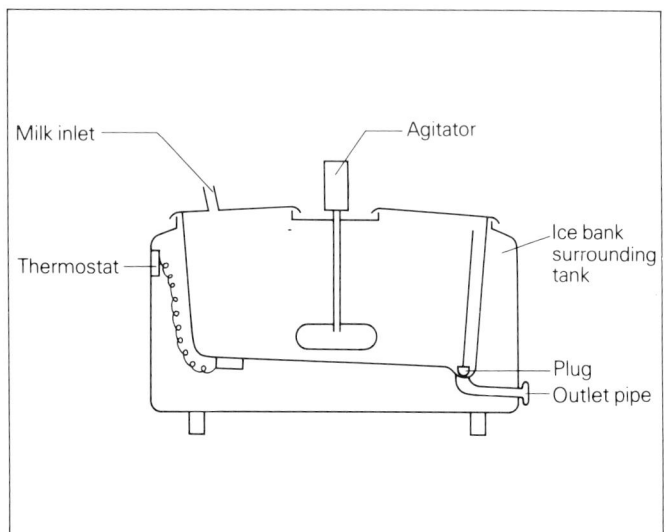

Figure 9.8 A bulk milk tank

Figure 9.9 A handling facility for cows including a footbath and crush

For large herds producing over 4 500 litres/day a plate-cooling system which pre-cools the milk prior to transfer to an insulated partially-refrigerated bulk tank, is a viable alternative to conventional refrigerated bulk tanks.

Handling facilities

A planned system to handle cattle is essential for dairy units

Figure 9.10 Stalls for artificial insemination and veterinary treatment

with a low labour input. This is not only beneficial in reducing stress on the stock, but also to the stockman, AI inseminator, veterinary surgeon and others involved in handling the animals.

Figure 9.11 Self-locking yokes in a feeding passage

The handling facilities should preferably be sited adjacent to the milking parlour and incorporate a gathering pen, a forcing funnel, a race incorporating a footbath, a crush, and a shedding gate into two collecting yards (Fig. 9.9).

For artificial insemination it is often advantageous to have separate AI stalls (Fig. 9.10). It is important that the inseminator stands at the same level as the cow, so there should not be a step into the stall. A typical length should be 2.28 m long by 0.65 m wide. To separate cows at milking time for subsequent treatment or AI, a diversion gate at the exit from the parlour into a yard or box is usually beneficial.

An alternative handling facility for both AI and for many veterinary treatments is the provision of self-locking yokes in a feeding passage (Fig. 9.11). This allows all the cattle to be yoked daily and allows easy access to the head and rear of the cow.

Calving and isolation facilities

The most common type of calving accommodation is the calving box with an area of $16-25$ m^2. The problems with this type of arrangement are the lack of room to manoeuvre when assistance is required for a calving, and due to the difficulty in cleaning out such boxes there is often a build-up of disease giving rise to endometritis and coliform mastitis in cows, and navel and enteric diseases in calves.

The best indoor calving facilities are large straw pens which hold a number of calving cows, and which can be mechanically cleaned out. Alternatively, the provision of calving pens for individual cows which are easily dismantled and permit mechanical cleaning out can also be satisfactory.

It is rare that cattle have to be isolated from the herd to prevent the spread of disease. If this is necessary, the drainage, straw and dung have to be collected and separately disposed of, and the staff involved in feeding and treating the animals have to wear suitable protective clothing for cleaning and disinfection. For these reasons such isolation facilities are not normally found on farms, and calving accommodation is used for sick and injured animals.

Manure disposal

The move to cubicle housing systems has raised serious problems on many farms due to the large amounts of slurry (dung plus urine and washings) produced. Previously, housing systems dealt with manure in a solid form which could be stored either on concrete or on the land. Slurry, however, has to be contained and spread in such a way that it does not contaminate the watercourses, streams and rivers.

Slurry production and storage

The amount of slurry produced by a cow varies according to the feeding system, her size and her level of performance, but averages 50–60 litres/day at 10–15 per cent DM. In addition washings from the parlour, bulk tank and collecting yard amount to 15–25 litres/cow. In a 180-day winter, therefore, the total slurry production per cow is likely to be in the range of 12 000–15 000 litres. This total amount may also be increased if rainwater is incorporated, such as from an outdoor self-feed silage system.

Alternative slurry collection and storage systems are outlined in Fig. 9.12. The tractor-scraped passage is the lowest cost alternative, followed by the automatic scraper and slats. In a new building, however, slatted passages between cubicle banks can be narrower than solid passages (2.0 m versus 2.5 m) and a narrower building can be used, thus offsetting some of the additional costs associated with slats. The benefits from slats are that the cows can be kept cleaner than with solid passages, and they allow the incorporation of a storage tank beneath the slats.

A range of farm-built and purpose built slurry stores are used, the choice mainly depending on the capital available, and whether it is desired to handle a semi-solid or liquid slurry.

Many farm-made storage compounds are part excavated and part made up of retaining earth bank sides. These are only emptied at infrequent intervals (12 months or more).

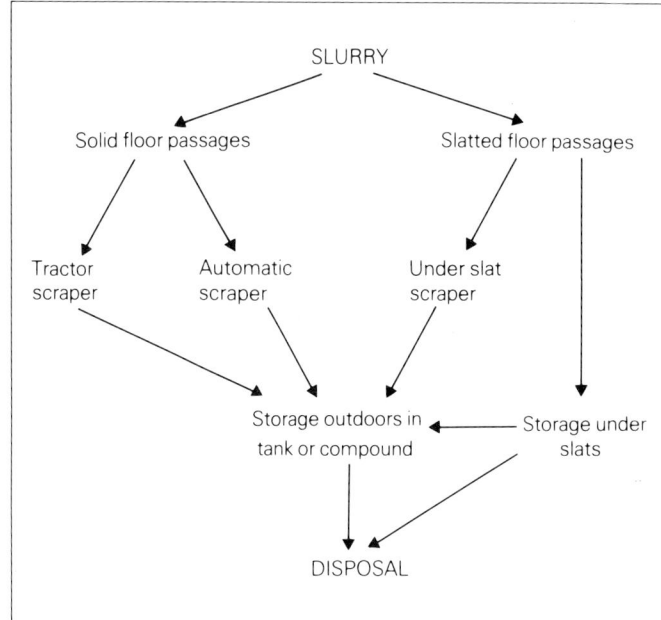

Figure 9.12 Alternative slurry transfer and storage systems

Other above ground types of smaller compounds are made of railway sleeper or concrete walls. It is common with many slurry compounds to have a section into which liquids are filtered. This separated liquid is pumped out either into irrigation pipes, or by vacuum tanker, and the remaining semi-solids in the main compound are handled and spread as solid manure.

Prefabricated above ground slurry stores have a concrete base with steel panel sides up to a height of 7 m. They are filled by a slurry pump, and emptied by gravity into a reception pit from which the slurry is pumped into spreaders (Fig. 9.13).

Slurry disposal

The weather and soil conditions mainly dictate when slurry can be spread on the land. In appropriate dry conditions slurry has to be spread quickly and satisfactory machinery is required.

Due to their small size (mostly up to 4 500 litres) and the need to manhandle the suction pipe into and out of the reception pit, the use of vacuum tankers is declining. The trend is towards tractor or electrically operated slurry pumps delivering into slurry spreaders of over 4 500 litres capacity to speed up disposal.

Treatment of slurry

The mechanical separation of slurry into solids and liquids is feasible and machines are available for this purpose. They

Figure 9.13 Slurry storage tank and tractor-operated pump for emptying from reception pit

are expensive and generally not economic unless, for example, the solids can be sold, such as for market gardening purposes.

Aerobic treatment of slurry involves the reliance on aerobic micro-organisms to break down the organic matter of slurry to produce carbon dioxide, water and a residual sludge. The treatment results in a product with a much reduced biological oxygen demand (BOD) which is much safer for disposal. This aerobic treatment also has the advantage of reducing the smell. The action takes place naturally in shallow storage compounds, and can be artificially induced in special oxidation ditches, by perforated pipes pumping air into the slurry and by various types of air diffusers.

The anaerobic breakdown of slurry is caused by the action of micro-organisms in the absence of oxygen. This takes place naturally and results in the production of hydrogen sulphide, methane, carbon dioxide and ammonia, leaving sludge and water. It is this type of degradation which gives rise to the foul smell of stored slurry. In specially designed plants the methane can be stored and used as a source of energy.

Nutrient content of slurry

Most farmers regard slurry primarily as an effluent disposal problem. However, it is a valuable source of nutrients which can produce a saving in fertilizer use.

Cow slurry with 10 per cent DM contains approximately 0.5 per cent nitrogen, 0.2 per cent phosphate and 0.5 per cent potash. Some losses of nitrogen occur during storage, varying with the duration of storage and the amount of agitation carried out, and range from 10 to 30 per cent. The nutrients contained in slurry are not all available in the year of application. When slurry is spread in the spring, about 50 per cent of the nitrogen and phosphate, and 90 per cent of the potash, are available in that season. The available nutrients are therefore about 2.5 kg nitrogen, 1.0 kg phosphate and 4.5 kg potash/1 000 litres of slurry.

Problems with slurry

Slurry, whether spread on land immediately or after a period of storage, does have an objectionable smell. This tends to be worse in disposal systems which force fine droplets of slurry into the air such as from irrigation pipes and slurry guns. The smell is less from most slurry spreaders unless they force the slurry into the air. On farms situated close to houses this problem has to be given consideration and the alternatives are to aerobically treat the slurry, to separate the solids and liquids, and in extreme cases to treat the slurry with a masking agent.

Pathogenic bacteria such as *Salmonella* species and *Escherichia coli* can survive in slurry for 2 to 3 months, and therefore slurry should not be spread on land to be grazed within that period. This is also important to prevent taint and subsequent rejection of the grass by grazing cattle.

The major problem arising from slurry is the pollution of watercourses, streams and rivers. This pollution leads to anaerobic conditions in the water and to the death of fish and other water creatures. Also there may be excessive growth of algae and weed plants in and around the water's edge. Slurry should therefore be spread when the soil moisture is low enough to prevent saturation and run-off. This presents a quandary to the farmer, because on heavy land the most satisfactory time to spread in winter is during frosty weather, and unfortunately this can lead to run-off into ditches and watercourses. It is essential that adequate storage capacity is available (at least 3 months) so that the farmer has maximum choice of spreading period.

High levels of slurry applied in single applications should be avoided. At levels over 40 000 litres/ha smothering and winter kill of the grass can occur. It is therefore preferable to apply two or three light dressings (15 000 to 25 000 litres/ha) during the winter and spring. An additional dressing can also be applied after a silage cut intended for subsequent cutting.

Labour management

On the majority of dairy farms all or most of the work is carried out by the family, and therefore labour problems are minimal. As herd sizes increase and a greater proportion of the work is carried out by employed staff, labour management increases in importance. Increasing herd sizes and mechanization means more cows to be looked after by each man, and increasing levels of cow performance means a greater need to do things the right way.

For the owner or manager it is necessary to employ good staff, to train and motivate them, to provide good working conditions and to maintain good relations.

Successful stockpersons are often found to have personalities which can be defined as confident introverts. They tend to prefer working alone for much of the day, understand all the facets of the job, and are capable of taking whatever action is necessary when things go wrong. The main ingredients therefore are a strong interest in the job, a perceptive nature and confidence in his or her own ability.

It is essential that staff should feel motivated if they are to carry out long hours of high quality work. The first requisite is to pay a good wage, and because the hours of work per day are often variable due to the seasonality of calving, it is normally preferable to pay a good basic wage all the year round, rather than a low basic wage plus overtime. The latter system only encourages inefficient working to maximize overtime. Many employers pay a bonus to dairy staff based on a range of elements such as milk output, live calves born per year, calving interval, margin over concentrates. If this kind of bonus system is used it is important that the factor on which the bonus is earned can be manipulated by the employee. To motivate staff successfully it is of course necessary to provide good working conditions and, if need be, good living conditions, and to make the employee feel a valuable part of the organization.

Both employer and employee must have a complete understanding of responsiblities, hours of work and pay, as misunderstandings can lead to dissatisfaction on both sides. The employee should also be given the necessary training in how to carry out jobs (such as work routines in the milking parlour), and where appropriate should be trained in new techniques (such as do-it-yourself AI) to give additional interest and responsibility.

The major aspect of employer/employee relations, however, is communication, and regular discussion on all aspects of the job, even if it results in few changes being made, has a great long-term benefit in job satisfaction on both sides.

Rearing replacements for the dairy herd

10

In the UK there is an annual requirement for about 0.75 million dairy heifers to replace cows culled from dairy herds.

Most dairy farmers choose to rear their own heifers for replacements as opposed to purchasing from the market or other farmers directly. The main reasons for this preference are: (a) that home-rearing helps to prevent the introduction of disease into the dairy herd; (b) that heifers can be trained and prepared nutritionally before entering the herd; and (c) most importantly, that it is the only satisfactory cost-effective method of genetic improvement. One other factor which should not be underestimated is that many dairy farmers would find it difficult to maintain enthusiasm for the dairy enterprise if they were not rearing their own replacements.

In herds where land or capital are very limited it is financially advantageous to have a 'flying herd' where all the land is stocked with dairy cows and the replacements are purchased. Even then, consideration should be given to either renting land for rearing, or contracting the rearing to another farmer in order to accrue the benefits of home-bred replacements.

Unfortunately on many dairy farms too much of the business resources (land, labour and capital) are devoted to rearing. A more disciplined approach involving rearing only the required number of heifer calves each year and calving heifers at a younger age will considerably increase the profitability of the dairy enterprise as a whole.

Number of heifers required

The culling rate in the herd determines the number of heifers to rear. Unfortunately many farmers rear all their heifer calves which leads to surplus heifers available and an enforced higher culling rate in the dairy herd. Unless there is a profitable market for calved heifers, it is preferable to rear only the minimum number of heifer calves.

For a 100-cow dairy herd with a 13-month calving index, the number of heifer calves produced annually is as illustrated in Table 10.1.

Table 10.1   Number of calving heifers produced annually in a 100-cow herd with a 13-month calving index

No. of calvings per annum	92
Heifer calves born (50%)	46
Heifers for bulling (assuming 5% loss at birth and 5% rearing losses)	41
Calving heifers available (assuming 7% fail to conceive and 3% rejected at calving)	37
Surplus heifers (if 25% culling rates)	12

In this example, if only 25 replacement heifers are required annually (the national average culling rate is about 25 per cent) it would only be necessary to serve 66 heifers and cows with a dairy-bred sire. The remaining 34 can be served with a beef bull to produce more saleable surplus calves.

If a satisfactory breeding programme is practised the heifers entering the herd are superior genetically to the cows, and it is beneficial to produce replacement calves from these heifers to speed up genetic progress. Also, because most heifers are served so that they calve at the beginning of the calving season (usually autumn) the female calves born from them can be reared to subsequently calve at 24 months old.

Age at first calving

The average age at first calving is about 33 months. This relatively late age is primarily due to low planes of nutrition during rearing giving rise to delayed puberty and first service. The rising costs of rearing encourage earlier calving, and in many countries, calving at 24 months is accepted practice.

A reduction in the age at first calving leads to:
1. A reduced land requirement for grazing and forage.
2. A reduced working capital requirement and interest charges.
3. A reduced labour, machinery and housing requirement.
4. A faster rate of genetic progress.

The area of land required for rearing declines considerably as the age at first calving is reduced (Table 10.2).

The land released from rearing can be used for more

Table 10.2 Area of land (ha) required for grazing and conservation according to age at calving and stocking rate

stocking rate* (livestock units/ha)	Age at first calving (years)		
	2	2.5	3
2.5	0.4	0.5	0.7
2.0	0.5	0.7	1.0
1.5	0.7	0.9	1.3

* One livestock unit is approximately equivalent to 1 calf + 1 heifer in a 2 year calving system (about 500 kg liveweight)

profitable purposes such as for keeping more dairy cows or for cropping. In evaluating the advantages of earlier calving one must therefore include the 'opportunity cost' of the released land. In reducing the age at first calving from 33 to 24 months, the interest charges and fixed costs are reduced by 30–40 per cent.

These direct financial benefits together with the faster genetic progress represent strong incentives for earlier calving.

The performance of these heifers on entering the dairy herd is, however, a further consideration. Results of a number of surveys have shown that although the first lactation yields of 24 month calving heifers are lower than for 36 month calving heifers due to their smaller size at calving, this advantage does not persist. Indeed, herd life tends to be greater for 24 month calvers compared with those calving at 30 or 36 months and their resulting lifetime yield is greater.

Extremely early calving at under 20 months of age associated with high growth rates (over 0.8 kg/day) should be avoided as this is known to lead to the laying down of fat in the udder, to a poor lactational performance throughout life, and very often to a poor reproductive performance.

In practice the age at first calving is determined by the calving pattern of the herd; often the autumn-born calves subsequently calve at 36 months and spring-born at 30 months. One simple way to overcome this is to calve the autumn-born calves at 24 months and serve the spring-calving cows with a beef bull. This will necessitate using a dairy bull on the heifers.

Liveweight targets

It is necessary in calving heifers at a younger age to achieve target liveweights at particular ages, otherwise the heifers will be too small on entering the dairy herd.

Puberty occurs at about 40 per cent of mature liveweight although at high planes of nutrition, puberty is reached at slightly lower weights than on low planes. Heifers should be served at about 15 months of age (13–17 months at a liveweight representing about 55 per cent of mature weight). Suggested target liveweights for different breeds are given in Table 10.3.

Table 10.3 Average liveweights and growth rates required for 2-year calving for different breeds of heifers

	Birth weight (kg)	Weight at puberty (kg)	Weight at conception (kg)	Weight before first calving (kg)	Weight gain birth to calving (kg/day)	Mature weight (kg)
Holstein	43	270	360	550	0.69	660
Friesian	40	250	325	510	0.64	600
Ayrshire	32	210	280	430	0.55	510
Guernsey	27	190	260	390	0.50	450
Jersey	24	160	220	340	0.43	380

General principles of rearing systems

The objectives in rearing dairy replacements should be to attain the target liveweights at the desired time with the least cost; to have low veterinary costs; for calf mortality rates to be below 5 per cent; and of the heifers served for 95 per cent to calve in the required calving period. To achieve these objectives requires a level of management equivalent to that put into the dairy herd.

Calf rearing systems

The major concern during the first few weeks of life is to maintain a low incidence of health problems. Also simple rearing systems giving a predictable level of performance are desirable.

Care after calving

When the calf is born the following should be adhered to:
1. The calving area should be clean and dry.
2. Remove mucus from nose and mouth.
3. Dress the navel cord with iodine solution or with aerosol antibiotic spray.
4. Ensure that the calf suckles the dam in the first 6 hours to give an adequate intake of immunoglobulins from the colostrum (minimum 3 kg of colostrum).
5. If calf does not suckle feed the colostrum in a bottle with teat or use a stomach tube.

Calves which have not received sufficient colostrum are liable to suffer from septicaemia in the first 14 days of life caused by *Escherichia coli* bacteria entering the bloodstream from the gut. This condition has a high mortality rate.

At about 24 hours after birth the calf should be removed from the dam and taken to a clean disinfected (preferably individual) pen. During the next 3 days the colostrum from the dam should be fed in a bucket or through a teat at a rate of 2–3 kg/feed, twice daily.

Management to weaning

At 5 days of age the liquid feed can be switched abruptly to milk substitute which is cheaper than wholemilk. A range of types are available including those based on whey and non-milk proteins, but the majority are based on dried skim milk plus added fat, minerals and vitamins. Some milk substitutes are also acidified with organic acids to maintain keeping quality after mixing and it is claimed that this gives additional protection against nutritional disorders.

Many complicated systems of feeding are practised but simple early weaning systems work just as well and are usually cheaper. Four such systems are outlined in Table 10.4, all involving early and abrupt weaning. During the first week after weaning the concentrate intake should increase to at least 1.25 kg/day.

An alternative feed is stored colostrum. The first 4 days' milkings can be stored in containers for a number of weeks. The colostrum ferments for 10–14 days producing a stable preserved material containing large amounts of lactic acid at a pH of 4.5 or less. The stored colostrum should be mixed before feeding and offered once or twice per day at a level of 3–4 kg or 1.5–2 kg/feed respectively. If shortages occur

Table 10.4   Systems of calf rearing on milk substitute (MS)

	MS feeding system			
	Twice daily bucket	Once daily bucket	Acidified teat	Machine teat
Penning	individual	individual	group	group
Age at start (days)	4–5	4–5	4–5	4–5
Age at weaning (days)	32–40	32–40	32–40	32–40
Amount MS powder per feed	250 g	500 g	1 kg/	1 kg/
Amount of water	2.5 litres	3 litres	7 litres*	7 litres*
Warm/cold feeding	Warm	Warm	Cold	Warm
MS powder usage (kg)	16–20	16–20	30–50	30–50
Approx LW gain (kg/day)	0.5	0.5	0.8	0.8

* ad lib

the feeding can be abruptly switched to milk substitute.

The calf concentrate can be pelleted or be a coarse mix with a crude protein content of not less than 16 per cent (fresh weight). This should be fed *ad libitum* until an intake of 1.5–3 kg/day is reached; the actual amount depending on the quality of the forage offered *ad libitum*.

Problems of septicaemia and scour   About 75 per cent of calf deaths occur during the first month of life, the major problems encountered being septicaemia and scour.

Septicaemia is caused by *E. coli* organisms entering the bloodstream and results from an inadequate colostrum intake during the first hours of life. Unfortunately the condition is often untreatable as sudden death occurs.

Scouring (diarrhoea) is normally caused by localized *E. coli* infections of the intestine. The faeces vary from white and pasty to a dark watery consistency. The diarrhoea results in the loss of electrolytes and water, and treatment of the disease should include dilution of the milk substitute and the feeding of electrolytes. For severe cases and other disease problems, the veterinary surgeon should be called to give relevant treatment.

Routine treatments

After birth the calves should be ear-tattooed or have a metal ear-tag inserted containing the herd letter and number, and the individual calf number.

The disbudding of calves should be carried out at 1–3 weeks of age using an electric cauterizing iron. A local anaesthetic must be given for this operation.

Supernumery teats should be removed at an early stage and it is usually convenient to do this at the same time as the disbudding operation. The extra teats are cut off with a sharp pair of sterile scissors, and then iodine is applied to the area.

Management from weaning to turn out

During this follow-on period the main objectives are to achieve an economic rate of liveweight gain of at least 0.65 kg/day, and to prevent the development of pneumonia in the calves.

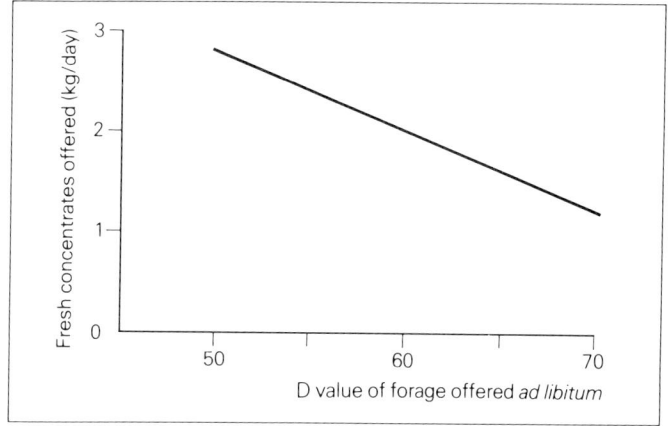

Figure 10.1 Amount of concentrates to offer with forages of different quality, to produce a liveweight gain of 0.75 kg/day

Feeding systems

The calves should be offered *ad libitum* uncontaminated forage together with a concentrate (at least 12.5 MJ/kg dry

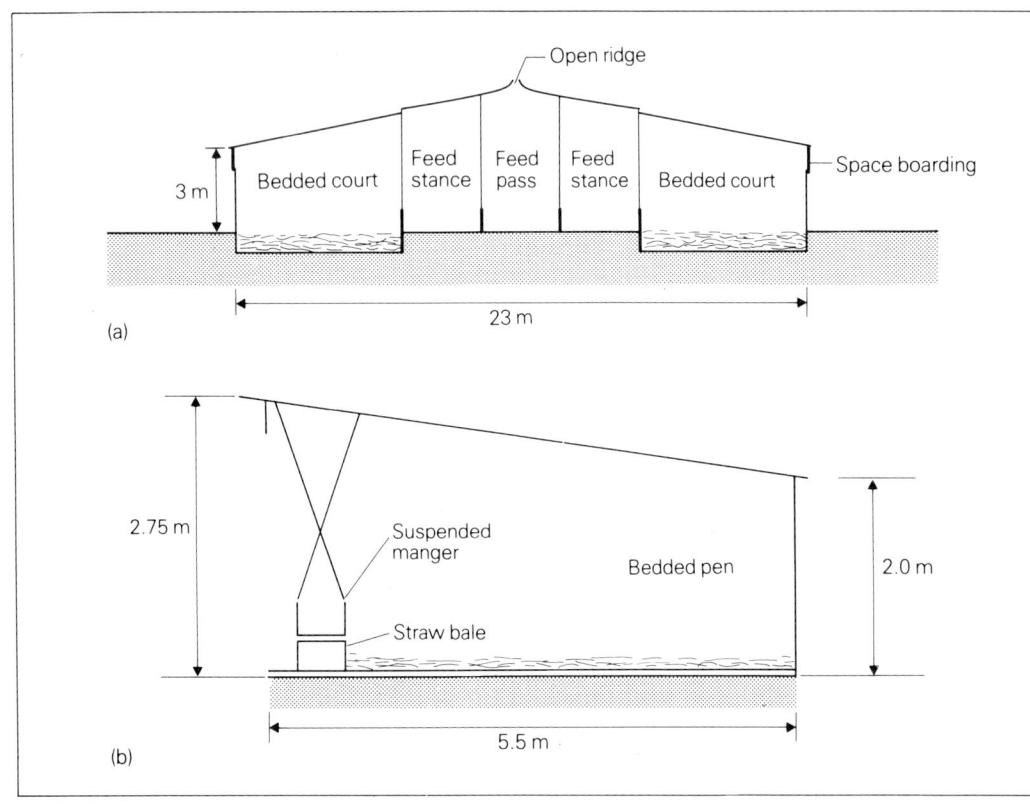

Open ridge

Bedded court | Feed stance | Feed pass | Feed stance | Bedded court

Space boarding

3 m

23 m

(a)

2.75 m

Suspended manger

Straw bale

Bedded pen

2.0 m

5.5 m

(b)

Figure 10.2 Examples of calf buildings with satisfactory ventilation: (a) Steel-framed, block-walled, asbestos roof calf building with bedded courts, concrete feed stance and centre feed pass. (b) Simple monopitch calf building with open front, constructed of block walls and asbestos roof

matter (DM) of metabolizable energy (ME) and 14 per cent crude protein fresh weight). The amount of concentrate to offer will be greater for lower quality forages as indicated in Fig. 10.1.

Prevention of pneumonia

The main scourge of calf rearing during this rearing phase is pneumonia brought about by poor environmental conditions in the calf house. The conditions that predispose to the disease normally include inadequate air space per calf due to a high stocking density with a low roof height and too little or too much ventilation. Examples of calf housing which have given a low incidence of pneumonia are illustrated in Fig. 10.2.

Management during the second winter

On grassland based dairy farms the objective should be to use silage as the basal indoor feed. This can be self-fed or easy-fed. If self-feeding is practised the silage should not be too compacted, otherwise the silage will be difficult to pull out by the heifers and poor intakes will ensue. A settled height of silage of no more than 1.8 m is preferable with a feed face width of not less than 0.15 m/head.

The amount of supplementation offered with forages

should depend on forage quality. For forages of over 10 ME (MJ/kg DM) silage fed in restricted amounts will produce the target growth rate of 0.6 kg/day (Table 10.5). If the ME value is less than 10 then concentrates will be required.

Table 10.5 Forage feeding of 350 kg heifers during their second winter (growth rate required 0.6 kg/day)

ME of forage (MJ/kg DM)	Forage intake (kg DM/day)	Concentrate intake (kg/day)
8	4.1	2.6†
9	5.5	1.2
10	6.0 (R)*	—
11	5.5 (R)	—

* (R) = restricted feeding
† 18% CP in DM

Heifers which have subsequently to be housed in cubicles, should be trained during this period and therefore cubicle housing for such heifers has considerable advantages. The dimensions should be approximately 2 m long by 1.05 m wide.

Management during the grazing season

The failure to achieve targets for earlier calving is often due to bad management during the summer months. These disappointing growth rates achieved over the grazing season are due to:

1. Low levels of grass production resulting from poor swards and inadequate fertilizer levels.
2. Understocking for the first 6 weeks of the season followed by overstocking thereafter.
3. Parasitic bronchitis (husk/lungworm).
4. Parasitic gastroenteritis (stomach worms).

Grass production and utilization

It is not necessary to have sown swards of grass to give high grass yields; well managed permanent pasture can be equally as productive. The level of nitrogen fertilizer

Table 10.6   Effect of nitrogen fertilizer level on the stocking rate of youngstock

Nitrogen fertilizer level (kg N/ha)	Annual stocking rate (livestock units/ha)*
0	0.50
75	1.15
150	1.75
225	2.25
300	2.60

* See Table 10.2

applied to youngstock pastures is often low, and one method of intensification is to increase fertilizer levels (Table 10.6).

Exploiting the grass production requires a matching of utilization with production throughout the grazing season. As the animals grow, their daily energy and protein requirements increase whilst at the same time grass production decreases. It is necessary therefore to have a declining stocking rate as the season progresses (Table 10.7)

Table 10.7   Changes in stocking rate (livestock units/ha)* necessary during the grazing season to match grass intake with grass production

Average stocking rate during the grazing season	Season		
	early	mid	late
5.0	6.6	5.0	3.4
4.0	5.3	4.0	2.7
3.2	4.2	3.2	2.2

* See Table 10.2

Parasitic infections at grass

Parasitic bronchitis is caused by the lungworm *Dictyocaulus viviparus*. The adult worms are found in the trachea and bronchi of the lung. The eggs laid by the female worms hatch quickly producing larvae which are coughed up into the mouth, swallowed and subsequently appear in the faeces. These larvae develop to the infective stage in as little as 5 days and then migrate onto the grass. The ingested larvae travel to the lungs via the lymph and blood after penetrating the intestinal wall.

The disease symptoms of coughing, bronchitis, pneumonia and loss of weight are seen 3–4 weeks after the larvae appear in the lungs. The severity of the disease is related to the rate of ingestion of the larvae, and at high rates of intake a high mortality rate can occur.

The most susceptible animals are calves in their first summer at grass. Most outbreaks occur between June and November. The disease can be treated with anthelmintics but the most satisfactory approach is to control the disease by vaccination before turning the calves out to grass. Two doses of an irradiated larval vaccine are given to calves over 2 months old at an interval of 4 weeks.

Parasitic gastroenteritis is generally caused by the abomasal nematode *Ostertagia ostertagi*. It also is a problem of calves in their first grazing season. The life cycle of the worm is illustrated in Fig. 10.3.

The disease is characterized by loss of weight and diarrhoea, and is commonly seen from July to October (Type 1). It affects a high proportion of the stock, although mortality is generally low. A second form of the disease also occurs (Type II) in late winter/spring. This results from

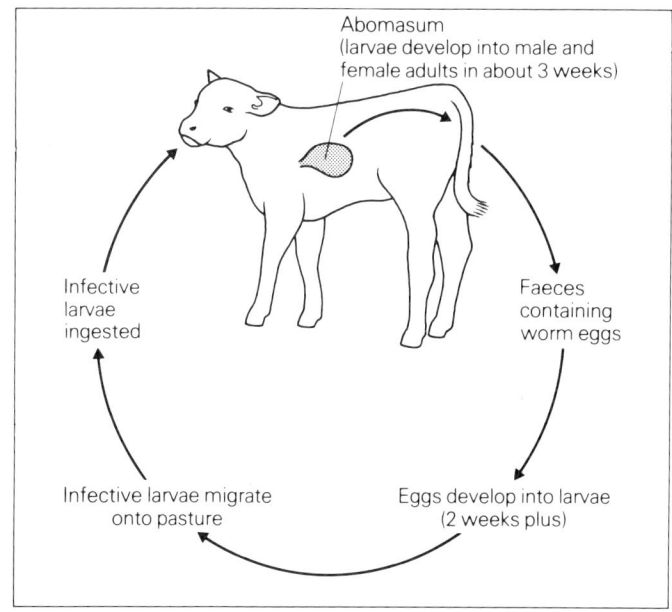

Figure 10.3 The life-cycle of the stomach worm (*Ostertagia ostertagi*)

larvae ingested in the late summer/autumn becoming inhibited in the abomasal wall. They subsequently develop into adult worms in the late winter/spring producing some similar symptoms to Type I, including weight loss and diarrhoea, but with a high mortality rate in those stock affected.

A number of anthelmintic drugs given as drenches, boluses and injections are now available to control both Type I and Type II ostertagiasis, although vaccination is not possible. Systems of grassland management can, however, be used to control the disease. The important factors to consider in developing systems of control are:

1. There is some overwintering of larvae on the grass.
2. Development of the egg to the infective larval stage on the sward is slow in the spring but speeds up with time; consequently there is a massive appearance of these larvae on the sward from July onwards, particularly in wet conditions.
3. The adult worm population in the abomasum is continually changing due to new adults developing from the larval stage and old adults passing out; the severity of the Type I disease thus depends on the number of larvae ingested per day.
4. The Type II disease is difficult to control by grazing management except by the use of clean aftermaths from August onwards –anthelmintic treatment at housing or of affected animals in the late winter/spring may therefore be necessary.

Grazing systems

A successful grazing system for dairy youngstock should provide a continuous supply of high-quality grass throughout the season, and prevent the build-up of stomach worms, so that growth rates of calves and heifers are in excess of 0.7 kg/day.

Two such systems are outlined in Fig. 10.4. The leader – follower system developed at the National Institute for Research in Dairying (NIRD) allows the calves in their first grazing season to graze the paddock immediately ahead of heifers in their second grazing season. This system gives higher growth rates than separate grazing systems for calves and heifers, and also prevents the onset of stomach worm problems. The 1–2–3 system developed by ICI is a set-stocking system which was devised for beef cattle, but can be adapted for dairy youngstock by splitting the two age-groups with a temporary electric fence. This system also gives excellent growth rates with little build-up of stomach worms. Both these systems control the build-up of stomach worms by maintaining a low concentration of larvae on the herbage. This results from the integration of grazing susceptible calves with older cattle and with conservation. Treating the calves with anthelmintic during the season is unlikely to be of benefit in most years if the systems are practised as described.

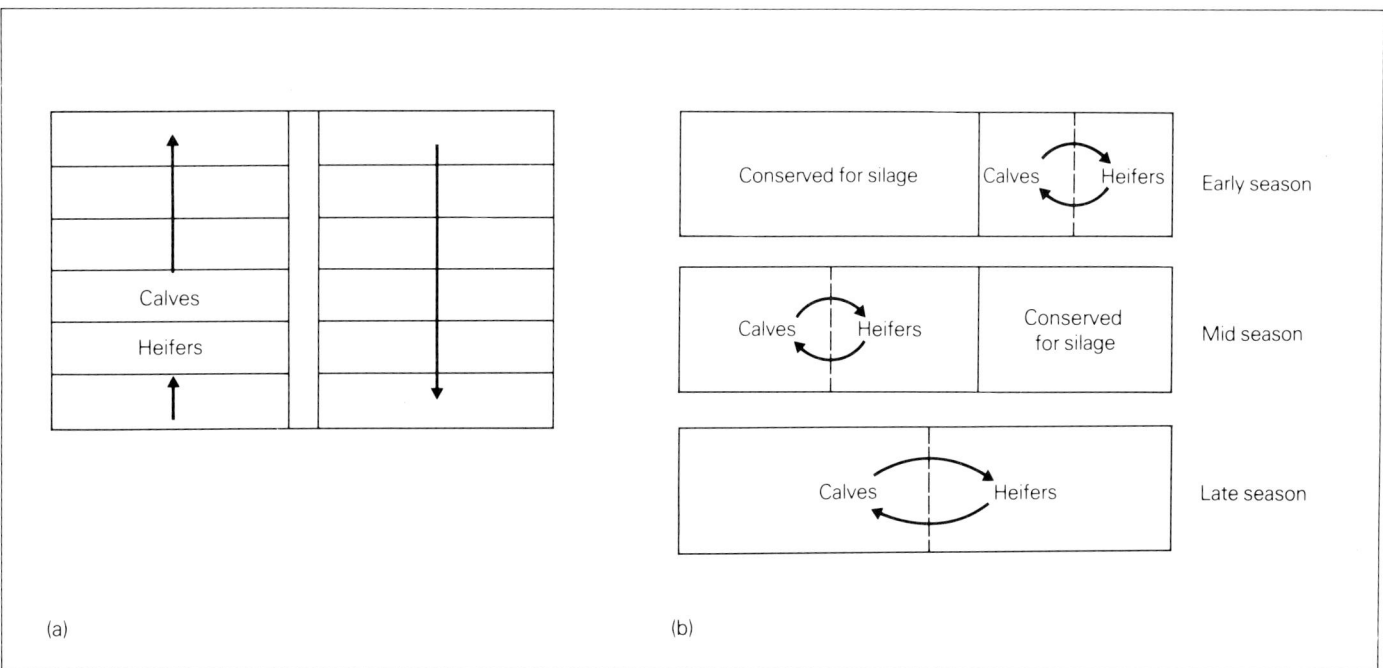

Figure 10.4 Grazing systems for dairy youngstock. (a) Leader/follower – calves graze paddock ahead of heifers. In spring about five paddocks are grazed and seven conserved, in mid season grazing and conservation are reversed, and in late season all are grazed. (b) 1–2–3 system – grazing and conservation are also integrated. The calves and heifers are set-stocked but separated by an electric fence, and switch fields at weekly or fortnightly intervals

Example rearing systems

Many systems of rearing are possible to achieve the target liveweights outlined earlier, and two such systems for autumn and spring-born calves, subsequently calving at 24 months are given in Table 10.8.

Whatever the system selected it is vitally important that there is a planned programme of management to attain the objectives, and a monitoring system to ensure that these are being met. The weighing of cattle is beneficial, but if this is not feasible the use of weighband gives a useful indication of the average weight of a group of animals, although there are inaccuracies with this method for individuals.

Breeding management

Choice of sire

Where possible heifers should be bred to dairy sires with a high ICC. If the breeding programme of the herd is progressive, then the heifers should have a higher genetic merit on average than the rest of the herd, and if the heifers are bred to beef bulls (as in many herds) then up to 30 per cent of potential genetic progress will be lost.

The main reasons for using beef bulls on dairy heifers are to reduce the incidence of calving difficulties (dystokia), and to remove the problems associated with using AI on heifers

Table 10.8 Examples of rearing systems for autumn-and spring-born calves subsequently calving at 24 months of age

Month	LW (kg)	LW gain (kg/day)	Concentrates	Bulk feed (unrestricted)
AUTUMN				
Sept.	40		14 kg milk sub. + 50 kg conc.	
		0.50		Hay
Nov.	70			
		0.65	350 kg conc.	Grass silage
April	180			
		0.70	100 kg barley	Grazing
Oct.	310			
		0.50	250 kg barley	Grass silage
April	400			
		0.70	—	Grazing
Sept.	510			
SPRING				
Feb.	40		14 kg milk sub. + 50 kg conc.	
		0.50		Hay
April	70			
		0.70	150 kg conc.	Grass silage
June	110			
		0.65	100 kg barley	Grazing
Oct.	190			
		0.60	270 kg barley	Grass silage
April	300			
		0.70	—	Grazing
Oct.	430			
		0.70	180 kg barley	Grass silage
Feb.	510			

(heat detection, regular sorting out of heifers for AI). By the use of good management, these problems can be overcome.

Dystokia in heifers

The incidence of dystokia in Friesian/Holstein heifers can be over 10 per cent when a Friesian/Holstein bull is used. The main cause of the problem is oversized calves relative to the size of the pelvic opening. Calves of over 40 kg birthweight or with birthweights which represent over 9 per cent of the postcalving weight of the heifer are likely to give problems. The conformation of the calf is also important.

There is a large variation between sires within breeds in the incidence of dystokia produced, and therefore selection of a high ICC sire for heifers which is known to produce a low incidence of dystokia is necessary.

High levels of feeding during the last 3 months of pregnancy can increase the incidence of dystokia due to the heifer laying down fat in the pelvic area. Also, high levels of feeding earlier in pregnancy may increase calf size. The aim should be therefore to have a steady rate of gain throughout pregnancy without resorting to periods of exceptionally high growth rates (over 0.8 kg/day).

Synchronization of oestrus

The use of synchronization techniques allows AI to be used in heifers without resorting to observations of oestrus. The most commonly used method is to give two prostaglandin injections 11 days apart followed by AIs at 72 hours and 96 hours after the second injection or by a single AI at 78 hours. An additional advantage of synchronization is that a high proportion of the heifers will calve close together at the desired time.

After the synchronized AI at which 50–75 per cent should conceive, the heifers which are not pregnant can either be observed (most returning 19–22 days after AI) and re-inseminated, or a beef bull can be run with the group to pick-up the returns.

Future attitudes

Most dairy farmers in the future are likely to have a dairy replacement enterprise on the farm due to their interest in breeding and as it is the most satisfactory method of improving the genetic merit of the herd.

The continuing squeeze on profits from dairying will necessitate a more positive approach to rearing, resulting in fewer replacements being reared combined with calving at younger ages than the present average of 33 months. This will mean a smaller proportion of land, labour and capital is devoted to the youngstock enterprise.

Targets in liveweight, mortality and fertility will therefore have to be increasingly achieved, requiring a high standard of management throughout the rearing period.

Tailpiece

The changes in systems of milk production which occurred during the 1960s and 1970s were dramatic. The national herd, whilst remaining relatively constant in size, improved considerably in genetic merit and was managed better, by a rapidly diminishing number of dairy farmers.

The trend to larger herd sizes with increasing milk yields per cow was brought about by application of the available scientific principles in breeding, feeding, grass and forage production, milking, housing and mechanization, and disease prevention. The main stimulus for these changes was the declining profitability of milk production per cow, as costs continued to rise at a faster rate than milk prices.

The successful dairy farmer has increasingly to view the farm business as a whole and to make decisions on how best to use the resources available to him (land, labour and capital). It may seem unfortunate that milk production is becoming more of a business and less of a way of life; however, whatever the rights or wrongs, this evolution is unlikely to be reversed. Much of the mystique associated with breeding better cattle, with feeding for higher milk yields and with stockmanship, is increasingly open to scientific evaluation, and as a consequence milk production is becoming more of a science and less of an art.

The traditional approach of giving attention to the individual cows which form the herd is gradually being replaced by systems of management which are geared to the efficiency, wellbeing and profitability of the herd as a whole. With this philosophy, the individual only becomes important when its condition or performance is very different from the rest of the herd, such as when it is bulling, sick, due to dry off, due to calve and so on.

An examination of the results of costed farms clearly illustrates that 'more cows per man' is no detriment to performance and economic efficiency; the milk yield per cow and gross margin per hectare increase on average with increasing herd size up to about 200 cows.

Whether these trends continue in the future will be determined by market forces, and in particular by market demands. If, as seems likely, surpluses of milk and dairy products become an increasing problem in the western world, then price constraints in some form are likely to be imposed by governments to prevent increased production. The objectives of the dairy farmer will then be geared more towards utilizing land, labour and capital to produce a given amount of milk in the cheapest way, as opposed to increasing the amount of milk produced per cow and per hectare as at present.

Research and development work has contributed substantially to these changes in the dairy industry. Its role in the future is unlikely to be less important. The industry will become increasingly competitive and better educated dairy farmers will become more demanding of new information on systems of milk production.

Further reading

The Federation of United Kingdom Milk Marketing Boards (1980) *Dairy Facts and Figures* 230 pp.

Etgen, W.M. and **Reaves, P.M.** (1978) *Dairy Cattle Feeding and Management*, John Wiley and Sons, New York, 638 pp.

Haresign, W. and **Cole D.J.A.** (1981) *Recent Developments in Ruminant Nutrition*, Butterworths, London, 367 pp.

Holmes, W. (1980) *Grass – its Production and Utilization*, The British Grassland Society, Blackwell Scientific Publications, Oxford, 295 pp.

Hunter R.H.F. (1982) *Reproduction of Farm Animals*, Longman, London and New York, 176 pp.

NIRD (1977) *Machine Milking*, Thiel, C.C. and Dodd, F.H. (eds), National Institute for Research in Dairying, Reading, 391 pp.

National Institute for Research in Dairying (1981) *Mastitis Control and Herd Management*, Bramley, A.J., Dodd, F.H. and Griffin, T.K. (eds), Technical Bulletin 4, The Hannah Research Institute, Ayr, 290 pp.

Swan, H. and **Broster, W.H.** (1976) *Principles of Cattle Production*, Butterworths, London, 438 pp.

Whittemore, C.T. (1980) *Lactation of the Dairy Cow*, Longman, London and New York, 94 pp.

Glossary of abbreviations

ABW	acidified boiling water
AI	artificial insemination
ACR	automatic cluster removal
BOD	biological oxygen demand
CP	crude protein
DCP	digestible crude protein
DE	digestible energy
D	digestible organic matter in dry matter
DM	dry matter
EBL	enzootic bovine leucosis
ECU	European Currency Unit
EEC	European Economic Community
GE	gross energy
GJ	gigajoule(s) (megajoule × 1000)
ICC	improved contemporary comparison
LW	liveweight
ME	metabolizable energy
MADF	modified acid detegent fibre
MOC	margin over concentrate (milk price over concentrate cost)
MJ	megajoule(s)
M/D	megajoules of metabolizable energy in the dry matter
NE	net energy
NFE	nitrogen-free-extractive
NIRD	National Institute for Research in Dairying
NPN	non-protein nitrogen

ODR	oestrus detection rate
PD	pregnancy diagnosis
RDP	rumen degradable protein
SNF	solids-not-fat
UDP	undegradable protein
VFA	volatile fatty acids

Index